Physics of Radiofrequency Capacitive Discharge

Physics of Radiofrequency Capacitive Discharge

V. P. SAVINOV

CISP

CRC Press
Taylor & Francis Group
Boca Raton London New York

CRC Press is an imprint of the
Taylor & Francis Group, an **informa** business

CRC Press
Taylor & Francis Group
6000 Broken Sound Parkway NW, Suite 300
Boca Raton, FL 33487-2742

First issued in paperback 2020

ISBN-13: 978-0-367-57162-7 (pbk)
ISBN-13: 978-1-138-60080-5 (hbk)

Visit the Taylor & Francis Web site at
http://www.taylorandfrancis.com

and the CRC Press Web site at
http://www.crcpress.com

Contents

This book is dedicated to Lenchen

Foreword

The current stage of scientific and technological progress is determined, first of all, by the need for accelerated development of high technologies, among which the leading positions are occupied by plasma technologies. As is known, the level of the latter is determined by the results of fundamental research in the physics of gas discharges, in particular, radiofrequency discharges (RF) of low and medium pressure.

The mention of even a small part of the list of already implemented or supposed applications of radiofrequency capacitive discharge (RFCD) shows its far from exhausted possibilities: the creation of a wide class of gas lasers, the development of 2- and 3-frequency technological plasma-chemical reactors, new generation plasma sources with an effectively controlled wide electronic energy spectrum (EES), the creation of RF sources of technological homogeneous plasma at high field frequencies (of the order of and above 100 MHz), light sources with a line spectrum for scientific purposes and practical applications, and others.

It should also be noted that along with numerous journal publications on RF discharge, there is a striking shortage of scientific monographs on this subject. Their list is very limited: two books contain short chapters on the basics of the physics of RFCD (Chapman B., Processes in a glow discharge – sputtering and plasma etching, New York, Wiley, 1980; Raizer Yu.P., Physics of gas discharge, 3rd ed., Dolgoprudny: The Intellekt Publishing House, 2009), the book by Raizer Yu.P., Shneyder M.N. and Yatsenko N.A. Radiofrequency capacitive discharge, Moscow, Nauka-Fizmatlit, 1995 and the recently published book Chabert P., Braithwaite N., Physics of Radio-Frequency Plasmas, Cambridge University Press, 2011.

This monograph, largely based on the original experimental research of the author, is devoted to the sequential presentation of low- and medium-pressure RFCD physics.

Particular attention is paid to the experimentally observed physical phenomenon of the appearance of high-energy near-electrode electron beams in the RFCD. Moreover, the novelty of the results obtained is due to the study of RFCDs with dense near-electrode electron beams under the conditions of a sharply asymmetric discharge with a small area of the active electrode.

The most important application of the RFCD studied is the possibility of using a near-electrode electron beam as a tool for the formation of a given plasma EES and excitation of beam-plasma instabilities in a gas-discharge plasma (BPI).

The author expresses his deep gratitude to Prof. Henri Amvrosievich Rukhadze for his attention and mobilizing support, he always remembers with gratitude Prof. Anatolii Alexandrovich Kuzovnikov for his great all-round assistance in all matters, sincerely thanks Valentin Anatol'evich Ryabova for many years of cooperation in the scientific and technological fields, his colleagues-pupils Valery Georgievich Yakunin, Vladimir Leonidovich Kovalevsky, Igor Fedorovich Singayevsky for all conducted joint research, Maxim Sergeyevich Kruglov for fruitful effective cooperation under my leadership recently, as well as Yuri Vladimirovich Savinov for invaluable assistance in preparing the monograph for the press.

Introduction

In modern technologies, requirements are constantly increasing on the parameters of the technological environment. Often, to solve scientific and practical problems in the optimal regime, it is necessary to create a strongly non-equilibrium plasma with a given character of non-equilibrium. In this connection, the problem of creating a plasma with well-regulated parameters, in particular, with a precisely controlled electron energy spectrum of the plasma (EES), is acute. Obviously, this can be achieved only on the basis of a detailed study of the physical mechanism of numerous gas-discharge processes and phenomena.

At present, it can be considered generally accepted that the problem of radiofrequency discharge (RFD), especially low- and medium-pressure, is the most interesting in the fields of physics of gas discharges and gas electronics and for the most important practical applications.

For decades, researchers have been paying much attention to RFCD, the most valuable feature of which is, in particular, the experimentally discovered possibility of very effective control of the EES of its plasma.

It suffices to mention only some of the most important practical applications of RFD, such as: 1) plasma technologies for surface modification (etching, deposition of thin films), 2) nanotechnology, 3) gas lasers, 4) plasma chemistry, 5) plasma displays, 6) exhaust gas purification systems for internal combustion engines, 7) switches of powerful power–electric systems, 8) ionic motors for space research, so that the urgency of further efforts to study this type of discharge is not in doubt.

The number of monographs issued for the entire history of studying RFD is only counted in units [1–6]. Quite recently a monograph dedicated to this subject was published [7].

Apparently, the above is explained by the complexity and multidimensionality of the very problem of the RFCD, and even by quite far from completing the process of its solution to date.

The attempt to write this book by the author, who for several decades was engaged in an experimental study of the physical mechanism of low-pressure RFCD, is mainly due to the need to create a scientific guide for a detailed acquaintance with the properties of this category, which is in high demand at the present time. The extensive experimental material, borrowed from the scientific periodical literature, and, basically, the original material, obtained in the process of research, conducted at the Department of Physical Electronics of the Lomonosov Moscow State University, can be of interest both for students, graduate students and experimental specialists, and for theorists who are trying to create physical models of discharge close to reality.

In accordance with the focus of the book – to explain to the maximum extent the physics of RFCD, the presentation is based on a detailed examination of the processes and phenomena that make up the physical mechanism of the discharge, with minimal use of the mathematical apparatus.

In this book, due to lack of space, purely technical issues related to the practice of obtaining RFCD, the electrical circuits of RF generators, the methods of their matching with the load, and others are not considered. Such useful information, although in a limited scope, can be found, for example, in [6, 8].

It seems useful to give a brief historical overview of the development of scientific research in the field of physics of radiofrequency discharge, highlighting, although rather conditionally, a number of time steps.

The first stage (1884–1950) begins with the experimental discovery by W. Hittorf in 1884 of the appearance of a glow of the residual gas in a vacuum tube placed in an inductor through which a high-frequency current is transmitted [9]. Then N. Tesla [10], D.D. Thomson [11] and other researchers pointed out this phenomenon, showing the possibility of the existence of two types of discharge, which are caused by the electric and magnetic radiofrequency fields of the inductor windings. After 1920, a somewhat more systematic study of the low-pressure discharge in the radiofrequency range begins. At the same time, the focus was on the phenomenon of radiofrequency breakdown of the discharge gap, the dependence of the ignition voltage of the discharge on the pressure and type of gas, the frequency of the RF field, the interelectrode distance, and the configuration of the electrodes. The study of the properties of a stationary RFD has received little attention, occasional studies have been devoted to measuring the rms electric field strength and comparing it with the field strength in a positive column plasma in a glow discharge of a direct current (DCGD), comparing the average electron energies in the RFCD and DCGD plasma, and studies of the spatial structure of the glow of RFCD. Investigations were carried out to develop methods for diagnosing the plasma of the RFCD – method of the Langmuir probe [12–14], the double probe method [15, 16], and spectroscopic measurements [17–19]. First attempts were made to explain the mechanism of the RFCD theoretically, based on the consideration of the motion of a free electron in the RF field in a gas [11]. The results of investigations obtained during this period were discussed in a number of review papers [2, 20–22]. The second stage (1950–1965) is characterized by the expansion of the research front, as well as the appearance, albeit very simplistic, but productive physical models of RFCD [23]. The study of the nature of the motion of electrons in radiofrequency fields was continued with different ratios of the frequencies of the RF field and collisions of electrons with neutral particles, the mechanism of energy collection by electrons arising in inhomogeneous RF fields of time-averaged forces acting on charged particles [2, 24, 25]. Until the early 1960s, the mechanism of radiofrequency breakdown was mainly investigated [2, 20]. Substantial attention was paid to measurements of quasi-stationary electric fields in the discharge gap [23, 26, 27], and the mechanism of their appearance in the near-electrode regions was explained [28]. Spatial distributions of the electron concentration and temperature, the RF field intensity, the local plasma potential, the intensity of the integral luminescence, and individual spectral lines were studied [27, 29–35]. Experiments were carried out on the characteristics of a radiofrequency discharge such as the current–voltage characteristic (CVC) [34], the plasma parameters [27, 29, 31, 33], the RF field intensity [31, 32], the energy of electrons [35] and ions [23, 36] on discharge electrodes, and separate measurements of the function of the energy distribution of electrons in the RFCD plasma were conducted [37].

The calculations of the RFCD characteristics began to be saturated with details of the real physical discharge mechanism [23, 25, 30, 38, 39]. A radio-technical representation of a capacitive RF discharge in the form of an equivalent electric circuit was proposed [30, 39]. Here we should especially note the experimental work of S.M. Levitsky and his proposed qualitative model of a capacitive radiofrequency discharge with two modes (α-discharge and γ-discharge) [23], which had a significant impact on the concept of this type of discharge for more than 30 years. During the third stage (1965–1975), the accumulation of experimental data on low-pressure RFCDs continued. The technique of probe measurements [40] was worked out, including the determination of the electron energy distribution function in the centre of the RFCD [41], a comparative study of the plasma parameters of the 'positive column' of RFCD and DCGD [42] was carried out, the optical emission spectrum of the plasma was investigated [43, 44], and the quasi-stationary electric fields and currents were measured in RFCD [45, 46]. A series of systematic experimental studies of the physical mechanism and parameters of low-frequency RFCD was carried out at the Moscow University, the main result of which was the discovery of near-electrode high-energy electron beams and the near-electrode plasma EES anomalously enriched with high-energy electrons [47–51]. The presence of electrons with energies above 1 keV in the plasma of the RFCD is confirmed by the data of other authors [52, 53]. The research began of the physical processes in the RFCD within one period of the RF field [51, 57].

The next stage, stage IV (1975–2000), is characterized by the intensification of scientific research, stimulated by the need to create high technologies, improving diagnostic methods, building RFD models with extensive computer-assisted methods. Only individual papers are devoted to the study of RF breakdown [58]. The focus is on the stationary RFD, the methods of optimal RF power insertion into the gas-discharge plasma [59, 60] and systematic measurements of the macroparameters of the RF current–voltage characteristics, the impedances of various discharge regions, the drops of the constant and alternating electric fields in the discharge, the active and reactive components of the discharge current, absorbed plasma of RF power, spatial structure of the discharge [61–69]. Particular attention was paid to the study of near-boundary layers of space charge in the RFCD: their parameters and spatial structure [70–82], charged particle fluxes at electrodes [83–88], and time-resolved measurements of charged particle energies were carried out [89]. The energy fluxes from the electrodes to the discharge gap and the influence of the NESCL on the properties of the near-electrode plasma were also studied [90–93]. Electron-emission processes at electrodes were investigated [94]. Considerable attention was paid to the investigation of elementary processes [84, 95–102] and electron kinetics in the RFCD plasma [67, 102–120]. In the literature, there are a number of models of RFCDs, usually assuming numerical simulation [67, 120–128]. A large number of studies pursued specific applications (plasma technology, plasma chemistry, laser technology, ion sources, etc.) [6, 129–144]. A large series of experimental studies devoted to the study of boundary effects and the role of near-electrode electron beams in the RFCD was carried out [145].

Finally, the present stage (2000 up to the present), in connection with the need to advance the field to higher frequencies ($f > 13.56$ MHz), is distinguished by a fundamentally new moment: the need to take into account both in theory and in experimental studies potential and vortex component of the electric field [5, 146] and the formulation of a more precise physical principle of the separation of RFD into E- and H-type discharges [147]. In the traditional frequency range ($f \leq 13.56$ MHz), the study of poorly investigated discharges remains relevant: 1) high-pressure RFCD ($p \sim 1$ atm) and 2) radiofrequency inductive discharge (RFID) of low and medium pressure ($10^{-3} \leq p \leq 10^2$ Torr), with mandatory consideration in the second case of elements of the E-type discharge mechanism. There is a relapse of interest in studying the phenomenon of radiofrequency breakdown in gases, resulting in new data on the breakdown mechanism and the characteristics of the motion of electrons in strong electric fields [148, 149]. The study of RFCD with dense near-electrode electron beams in which excitation of beam-plasma instabilities is possible started [145, 169]. The physical properties of RFCD in the presence of the processes of formation and growth of nanoparticles in its plasma have been experimentally and theoretically investigated [295–297]. Recently two-frequency [150, 292, 293] and three-frequency [294] RFCDs have been studied to optimize practical applications, allowing independent control over the parameters of the bulk plasma and near-electrode layers. The flow of publications devoted to technological applications of RFCD, in particular, for the active development of plasma nanotechnologies, is continuously expanding.

Despite the age-old history of research, the study of RFD, both from a purely scientific and, in particular, practical points of view, remains very relevant.

The scientific interest is due to the fact that the RF discharge is a complex physical phenomenon that includes virtually all processes that are the essence of the physics of gas discharges and gas-discharge plasma, such as elementary processes in the plasma volume, electron emission processes at the electrodes, a complex of near-electrode phenomena, the formation of a strongly non-equilibrium plasma, the excitation of plasma instabilities, and others.

It should be noted that the vision of the physics of the RF discharge in the process of research undergoes a certain evolution. Initially, the RF discharge was represented as existing in the form of two phenomena – radiofrequency capacitive discharge and radiofrequency inductive discharge.

According to [147], the principle of dividing RF discharges into capacitive and inductive ones is based on the difference in the nature of the electromagnetic field in the discharge plasma. If the field in the plasma with a high degree of accuracy is potential and its vortex part is small, the discharge should be considered capacitive. If, on the other hand, the field in the plasma is almost purely vortical, then such a discharge is called inductive.

For a long time, independently, both types of RF discharge - RFCD and RFID, differing in the configuration of the conductors delivering RF power to the spatial domain of the discharge existence, were studied. In the first case, the discharge burns between the plates of the flat capacitor, and in the second – inside the inductor. Traditionally, as a rule, low frequencies of the RF field were used, which did not exceed the modern industrial frequency $f = 13.56$ MHz in the order of magnitude. Therefore, the RF field in RFCD was considered to be purely potential, and in RFID it was purely vortical. The limited nature of this approach to solving the general physical problem of the RF discharge has become visible for two reasons. First, even at lower field frequencies in RFID, the presence of a potential electric field was detected due to the unavoidable existence of parasitic capacitances in the discharge. Secondly, when using higher frequencies ($f \gg 13.56$ MHz), even in the geometry of RFCD, in accordance with the theory of the Maxwell electromagnetic field, an appreciable vortex component of the field appears. Thus, in the latter case, the discharge, regardless of the configuration of the electrodes, becomes a 'hybrid' RF discharge (HRFD) supported by both the vortex and potential components of the electric field. HRFD can be artificially created at low frequencies, using a discharge gap with two channels of RF power supply – a parallel connected inductor and a system of flat electrodes fed by one RF generator.

Since RFCD and RFID have their advantages in the creation of plasma sources, it is of considerable interest to study the physical properties of HRFD that preserve the basic elements of the mechanisms of both of these discharges.

A detailed study of the physical mechanism of RFCD allows creating highly informative methods for diagnosing both the characteristics of the discharge itself and the working areas of the process equipment. For example, the Department of Physical Electronics of the Lomonosov Moscow State University developed a non-contact method for measuring the parameters of the near-electrode space charge layer (NESCL) – its width d_s, the jump in the quasi-stationary electric potential in the layer U_s, the ion flux densities at the electrode j_{is} and of the charges n_{es} at the boundary of the layer, and also a spectroscopic method for measuring the density n_{e0} of the electrons emitted from the electrode surface and the Townsend coefficient γ of surface ionization.

Discharges with a variable electric field are realized in a very wide frequency range $0 < f \leq 10^{14}$ Hz. The place occupied by the RF discharge among them is shown in the following table:

Discharges at alternating current $0 < f \leq 10^5$ Hz
Radiofrequency) discharge $10^5 \leq f \leq 10^9$ Hz
Microwave discharge $10^9 \leq f \leq 10^{13}$ Hz
Optical discharge $f \geq 10^{13}$ Hz.

Before a detailed consideration of the specific physical properties of RFCD, we first recall the general provisions regarding the gas discharge as a physical phenomenon of the passage of an electric current in a gas. As noted by V.L. Granovskii [151], an independent current in a gas has a number of significant features that sharply distinguish it from the current in ordinary conductors:

1) the current in the gas develops only when the electric field reaches a certain minimum intensity (gas breakdown, ignition of the discharge), which is not present in conventional conductors; for dielectrics (solid, liquid), the breakdown process exists, but it can not lead to a long-standing current in the same medium – a powerful breakdown in the dielectric terminates in a gas channel in a pierced dielectric;

2) when a current flows through a gas, a variety of elementary processes occur both in the gas itself and on the surfaces of the bodies that bound it; therefore, an independent current in the gas is accompanied by a large number of phenomena, in addition to the magnetic field inherent in any current and thermal action, light, chemical and mechanical effects are also observed here;

3) the current in the gas is spatially inhomogeneous; in most cases the gas region through which it flows dissolves into a number of parts that differ from one another in electrical characteristics (field strength E, charge density n_e) and appearance (intensity, emission spectrum), which is due to the predominance of various elementary processes in different regions of the discharge;

4) the boundary conditions play an important role not only in the quantitative characteristics of the current, but also in determining the type of its shape; especially the value of the surface of the negative electrode, for example, heating to a higher temperature can lead to the transition of a direct curennt glow discharge (DCGD) into an arc discharge;

5) the main characteristic of the current in the gas – the current-voltage characteristic – differs sharply from the current–voltage characteristic in ordinary conductors ($I = U/R$ – Ohm's law); the independent current in the gas in general does not have such a single functional dependence $I = F(U)$, even a given discharge gap has different CVC depending on the type of current (arc, DCGD, Townsend discharge); in contrast to the constant conductors, in order to establish a general relationship between the current I and the voltage U, covering all possible processes in the gas, it is necessary to write down a system of integral and differential equations;

6) the basic forms of stored energy in the gas are completely different than those of ordinary conductors.

It follows that the independent current in the gas is indisputably the most complex form of electric current in matter.

A distinctive feature of the gas discharge is the extraordinary diversity of its forms, realized in various physical conditions. We will mainly consider the discharge of a reduced pressure in the range of $10^{-3} \le p \le 10^1$ Torr.

This book is devoted to the physics of RFD, supported by electromagnetic fields with frequencies $f \le 13.56$ MHz, when the RFCD is well described in the quasi-electrostatic approximation, practically in the absence of vortex fields. This type of discharge is well studied, although it is not dealt with in special monographs, with the exception of one [6], describing only the RFD, and continues to attract interest both from the scientific point of view and in relation to numerous practical applications. The extensive list of literature cited here does not at all pretend to be exhaustive. On the one hand, it is very difficult to do technically, and on the other hand it does not make sense, since it would inevitably lead to the spreading of the main subject of the examination and deviation from the main goal – to maximally highlight all the fundamental components of the physical mechanism of the RFD, a clear understanding of which will allow the creation of sources of plasma with given parameters.

Units of measurement of physical quantities used in the book are traditional for gas discharge physics: electrical quantities – in volts, amperes, ohms; the energy of the particles that make up the plasma – in electron volts, and their temperatures - in kelvins or electron volts; gas pressure in Torr.

For the convenience of the reader, the list of abbreviations used to reduce the amount of text is placed below.

Abbreviations

ARFCD –	asymmetric RF capacitive discharge
AE –	active electrode
CVC –	current–voltage characteristic
RF –	radiofrequency
RFCD –	radiofrequency capacitive discharge
RFCD 1 –	single-electrode RF capacitive discharge
RFCD 2 –	two-electrode RF capacitive discharge
RFID –	radiofrequency inductive discharge
RFD –	radiofrequency discharge
GE –	grounded electrode
LIF –	laser-induced fluorescence
NM –	normal mode
NG –	negative glow
TRFCD –	transverse RF capacitive discharge
BPI –	beam–plasma instability
PC –	positive column
NESCL –	the near-electrode space charge layer
PCDCD –	positive DC discharge column
NEB –	near-electrode electron beam
MW –	microwave
SRFCD –	symmetrical RF capacitive discharge
SCL –	space charge layer
WEB –	wall electron beam
TCGD –	DC glow discharge
FDS –	Faraday Dark Space
EES –	electronic energy spectrum
DCGD –	direct current glow discharge
BPD –	beam plasma discharge
SS –	spatial structure
RFG –	radiofrequency generator

Main physical properties and characteristics of radiofrequency capacitive discharge

1.1. General electromagnetic properties and classification of radiofrequency discharge

The variety of forms and modes of existence of the radiofrequency discharge (RFD) requires a preliminary description of the general electromagnetic properties and classification according to specific physical principles of varieties of this type of discharge in order to clearly define the subject of this book.

First of all, it is necessary to give an unambiguous definition of the concept of 'RF discharge', also known in the foreign scientific literature as 'radiofrequency discharge'.

In contrast to superhigh-frequency (microwave) and optical discharges, the RFD is considered a discharge in which, in particular, the following condition is satisfied

$$\frac{2\pi c}{\omega} = \lambda \gg d, \tag{1.1}$$

where $\omega = 2\pi/T = 2\pi f$ is the angular frequency of the field, T is the period of oscillations of the RF field, f is the cyclic frequency, λ is the wavelength of the discharge exciting the electromagnetic field, and d is the characteristic geometric dimension of the discharge gap.

By condition (1.1), the discharge is in the near zone of the field source.

The main point of classification of RF discharges is their division into two types: 1) capacitive RFD and 2) inductive RFD.

According to [147], it is proposed to use the natural principle of dividing the RFD into capacitive and inductive, based on taking into account the nature of the electromagnetic field in the gas-discharge plasma. If the field in the plasma is potential and its vortex component is small, the discharge should be considered capacitive. When the field in the plasma is almost purely vortex, this discharge is inductive.

In the literature [152, 153] there are also the inductive RFD '*H*-discharges', and the capacitive RFD '*E*-discharges', according to the nature of the field that determines the physical mechanism of maintaining the discharge. Recently, in addition, in works on physics of RF physics [5, 147] the potential electric field is called 'capacitive', and the vortex electric field is called 'inductive'

It is possible to experimentally create both of the above types of discharge by placing a discharge tube in the inductor L or the capacitor C of the resonant RF oscillatory circuit, respectively (Fig. 1.1). In the first case, the RF current flowing in the inductor L creates a vortex magnetic field in it, inducing a supporting discharge vortex electric field, the power lines of which are coaxial with coils of the inductor coil. In the second case, the discharge is caused by the potential electric field of capacitor C.

It should be noted, however, that the formal separation of all RFDs into two types (inductive and capacitive) is an essential simplification of the properties of real discharges. The fact is that often in practice a specific RFD includes elements of physical mechanisms of both capacitive and inductive discharges.

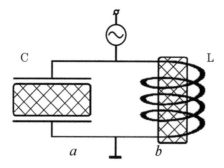

Fig. 1.1. Experimental scheme for creating radiofrequency capacitive and radiofrequency inductive discharges with an RF oscillatory circuit with a lower resonance frequency ($f \leq 13.56$ MHz).

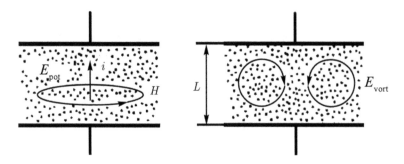

Fig. 1.2. Variable electric and magnetic fields in RFCD: *a*) geometry of the magnetic field lines; *b*) the lines of the vortex electric field.

As was already established in the first papers [11, 154], devoted to inductive RFD, the latter arises only because of the high voltage of the potential electric field at the ends of the inductor. Similarly, under certain physical conditions, the contribution of the vortex electric field to the maintenance of RFCD can be significant when the inductor is completely absent [146].

The frequency of the change in the electric field ω that supports the RFD is its most important characteristic. Let us consider the case when an RF voltage $V_\sim$ is applied to the discharge gap, creating a potential electric field E_{pot}. In this case, the RF discharge current $I_\sim$ excites an alternating magnetic field $H_\sim$, which in turn induces a vortex electric field E_{vort} whose lines of force are shown in Fig. 1.2.

Let us analyze the character of the electromagnetic field in a discharge without an external magnetic field on the basis of Maxwell's equations:

$$\mathrm{rot}\,\mathbf{H} = \frac{4\pi}{c}\mathbf{j} + \frac{1}{c}\frac{\partial \mathbf{D}}{\partial t}, \tag{1.2}$$

$$\mathrm{div}\,\mathbf{D} = 4\pi\rho, \tag{1.3}$$

$$\mathrm{rot}\,\mathbf{E} = -\frac{1}{c}\frac{\partial \mathbf{H}}{\partial t}, \tag{1.4}$$

$$\mathrm{div}\,\mathbf{H} = 0, \tag{1.5}$$

where $\mathbf{j}$ is the current density, ρ is the density of free charges, $\mathbf{D} = \mathbf{E} + 4\pi\mathbf{P}$ is the vector of electrical induction, $\mathbf{P}$ is the electric polarization vector (dipole moment of unit volume). Here it is taken into account that the plasma is a non-magnetic medium [155], in which the magnetic induction $\mathbf{B}$ and the magnetic field strength $\mathbf{H}$ are equal ($\mathbf{B} \equiv \mathbf{H}$).

Equations (1.2) and (1.4) define vortex fields, and equation (1.3) the potential electric field. In the RFD plasma, the fields vary in time according to the harmonic law $\mathbf{E}(t) = \mathbf{E}_a e^{i\omega t}$, $\mathbf{H}(t) = \mathbf{H}_a e^{i\omega t}$, where $\mathbf{E}_a$, $\mathbf{H}_a$ are the amplitude values of the electric and magnetic field strengths. From here:

$$\frac{\partial \mathbf{D}}{\partial t} \sim \omega, \qquad \frac{\partial \mathbf{H}}{\partial t} \sim \omega.$$

The equations for the vortex fields (1.2), (1.4) contain terms with the factors ω/c, and equation (1.3) for the potential field does not have terms with factors $1/c$.

It follows that for small ratios ω/c or, in other words, at relatively low frequencies of the field ω, the components of the vortex field E_{vort} and H_{vort} are small. In such cases, the electromagnetic wave is converted into a practically potential electric field E_{pot}.

The situation mentioned above arises for the field frequencies $f \leq 13.56$ MHz, i.e., less than or of the order of the often used industrial frequency. At these low frequencies, this potential field has a quasi-electrostatic character.

In the monograph [6] it is shown that the criterion for estimating the degree of quasi-potentiality of the electric field in the RFCD is:

$$\frac{E_{vort}}{E} \approx \left(\frac{\Lambda}{\sqrt{2}\delta'} \right)^2, \tag{1.6}$$

here $\Lambda = R\sqrt{L/L+R}$ is the characteristic size of the discharge cylinder with radius R and length L; $\delta' = c/\sqrt{2\pi|\sigma'|\omega}$; $\sigma' = \sigma + i\omega\dfrac{\varepsilon}{4\pi}$; σ and ε are the conductivity and dielectric permittivity of the plasma, respectively.

The physical meaning of the criterion of the potentiality of the field (1.6) is understandable from consideration of two limiting cases:

1) Suppose that the conduction current in the discharge is much larger than the displacement current, $i_{cond} \gg i_{bias}$, which is satisfied at medium gas pressures, then $\sigma' \approx \sigma$ and $\delta' = \delta = c/\sqrt{2\pi\sigma\omega}$ – depth of the skin effect in the conductor. Hence: $E_{vort}/E \ll 1$, when the characteristic size of the discharge is much smaller than the depth of the skin effect.

2) The bias current greatly exceeds the conduction current, $i_{bias} \gg i_{cond}$, which is performed at low pressures. In this case, it

can be shown [6] that $\delta' = \sqrt{2}c / \omega_{Le} = \sqrt{2}\lambda_{cr} / 2\pi$, $\lambda_{cr} = 2\pi c / \omega_{Le}$, $\omega_{Le} = \sqrt{\dfrac{4\pi e^2 n_e}{m}}$ is the electronic plasma frequency. Then, according to the criterion (1.6), the field will be potential if $\Lambda \ll \lambda_{cr}$. It follows that the lower the electron concentration n_e the more stringent the condition of potentiality of the field is satisfied.

In pure form, the *E*- and *H*-type discharges can be created only at lower field frequencies ($f \leq 13.56$ MHz), using the experimental scheme shown in Fig. 1.1.

Obviously, at higher frequencies ($f \gg 13.56$ MHz) in the RF circuit, it is not possible to obtain a pure *E*- or *H*-discharge, since in the capacitor *C*, along with the potential component E_{pot}, the vortex component E_{vort} appears, and in the inductance *L*, in addition to the vortex component E_{vort}, the potential component E_{pot} also appears, due to a stronger manifestation of the parasitic capacitances of the gas-discharge system.

Summing up what was said above, it should be emphasized that the traditionally researched RFCD at frequencies $f \leq 13.56$ MHz is supported by a potential quasi-electrostatic electric field. As a rule, earlier in the literature, with the exception of the monograph [6], attention was not accentuated on this.

It is necessary to establish the frequency range of existence of RFCD, substantiating from the physical point of view the characteristic values of its lower ω_l and upper ω_u boundary frequencies:

$$\omega_l \leqslant \omega \leqslant \omega_u.$$

The founder of the Russian scientific school of electronics N.A. Kaptsov believed that in qualitative terms *radiofrequency* discharge should be termed at such minimum frequencies ω_l when the space volume charge in the discharge gap no longer has time to be reconstructed during one period of the RF field [1]. An equivalent quantitative condition is the requirement that the amplitude of the oscillations of the electrons *A* in the RF field in the gas-discharge plasma does not exceed half the distance *d* between the electrodes $d/2$ [24]:

$$A \leqslant d / 2. \tag{1.7}$$

In the collisional movement of the electron (mobility mode [4]):

$$A = v_{edr} \frac{T}{2},$$

where $v_{edr} = b_e E_{\sim}$ is the drift velocity of the electron, $E_{\sim}$ is the strength of the RF field, $b_e = e/mv_{en}$ is the electron mobility, v_{en} is the frequency of collisions of an electron with neutral particles, e, m is the charge and the mass of the electron; from condition (1.7) we obtain an expression for the estimated frequency ω_l:

$$\omega_l \geqslant \frac{2\pi e E_{\sim}}{mv_{en}d}. \tag{1.8}$$

Let us consider the question of the upper boundary frequency ω_u for RFCD. Naturally, at higher frequencies, the vortex magnetic and electric components of the electromagnetic wave field will already appear. In the monograph [156] the opinion was expressed that it is reasonable to limit the corresponding RFCD frequency to the value at which the electric field becomes essentially non-uniform along the electrodes due to the skin effect. The characteristic size of this non-uniformity

$$L_{sk} = \frac{c}{\omega}\left(1 + \frac{\omega_{Le}^2}{\omega v_{en}}\right)^{-1/2}. \tag{1.9}$$

Hence, to determine the upper frequency ω_u, we obtain the expression:

$$L_{sk} = \frac{c}{\omega_u}\left(1 + \frac{\omega_{Le}^2}{\omega_u v_{en}}\right)^{-1/2} \leqslant D_{el}, \tag{1.10}$$

where D_{el} is the diameter of the electrode. Often in practice, the value of $\dfrac{\omega_{Le}^2}{\omega_u v_{en}} \gg 1$, so in this case, we can get a simplified expression for evaluating frequency ω_u:

$$\omega_u \leqslant \frac{c^2 v_{en}}{\omega_{Le}^2 D_{el}^2}. \tag{1.11}$$

Obviously, according to what was said above, the lower boundary frequency ω_l is estimated from the potential component of the field, and the upper ω_u is estimated from the vortex component.

As a result, we obtain an expression for estimating the frequency range of the RFCD:

$$\frac{2\pi e E_{\sim}}{m v_{en} d} \leqslant \omega \leqslant \frac{c^2 v_{en}}{\omega_{Le}^2 D_{el}^2}. \tag{1.12}$$

The actual frequency range of RFCD in practice is:

$$0.1 \text{ MHz} \leqslant f \leqslant 10^3 \text{ MHz}.$$

In addition to the separation of RF discharges into inductive and capacitive ones, other methods for their classification according to the determining physical mechanism of the discharge are also introduced and are discussed below. In [147], attention is further paid to the fact that it is important to distinguish between volume and surface, resonant and non-resonant discharges, depending on whether the frequencies of the RF fields exciting the discharge coincide with the resonant frequencies of the bulk and surface waves of gas-discharge plasma or are far from them, whether they penetrate or not in the volume of the plasma, and what is the nature of the skinning of such fields.

At the conclusion of this section, we make a series of summary remarks.

We return to the initial condition (1.1)

$$\lambda \gg d,$$

which characterizes the RF discharge as such. In connection with what was said above, we note at once that the condition (1.1) does not at all mean that the field inside the discharge plasma should be considered homogeneous. A field in a plasma can be significantly decreased near its surface, although it is possible to burn the discharge in a sufficiently large volume. In addition, the frequency of the exciting field ω can be close to the eigenfrequencies of the natural waves in the plasma, which leads to the excitation of an inhomogeneous wave field.

After the frequency range of RFCD ($0.1 \leq f \leq 10^3$ MHz) was set, we note the important physical property of the considered discharge. According to the monograph [155], in isotropic plasma there are no intrinsic eigenwaves in the frequency region of interest to us, in the frequency range $\omega < \omega_{Le}$ both longitudinal and transverse electromagnetic waves are screened, that is, they do not penetrate deep into the plasma. This type of discharge will be called *skinned* or *surface*, emphasizing thus that the plasma is opaque to the source field.

Another important property of the RFCD under consideration is that it is supported by a potential quasi-electrostatic RF electric field.

Up to now, the overwhelming majority of studies of the RF discharge have been devoted to the discharge with the properties indicated above. Recall that in industrial RF-equipment, as a rule, the frequency of the field is 13.56 MHz. It is this type of discharge that finds the widest practical application.

New queries of technology and technology are related, in particular, to the need to increase the working frequency of the high-frequency field ($f \gg 13.56$ MHz) and increase the area of electrodes that must be in contact with a homogeneous plasma [146]. This requires the creation of a theory, taking into account a number of electromagnetic effects in the mechanism of the RF discharge (standing waves, skinning, competition of the E- and H-type discharge mechanisms in a single discharge gap). It is necessary to carry out experimental studies of such discharges. These tasks are already on the agenda.

1.2. The main differences in the physical properties of RFCD and DC discharges

Historically, the scientific field of the physics of gas discharges has been formed on the basis of the study of a direct current discharge (DCD) as the simplest kind of electric current in a gas. Therefore, the DCD has been studied much better than other discharges. At the same time, practically all possible gas-discharge processes are represented in the DCD, which, due to the above, is a convenient methodological basis for studying the physical properties of various gas discharges.

We note the main differences in the physical conditions that exist in RFCD and DCD:

– in contrast to the purely active discharge current of the DCD, the total discharge current of the RFCD consists of active and reactive components, which causes the appearance of a number of new physical properties (for example, RFCD with external electrodes is possible);

– in the RFD, the electric field periodically changes its direction, as a result, in comparison with the DCD, the escape of charged particles from the discharge gap decreases substantially, and only part of the electrons emitted from the electrodes supports the discharge;

– the role of collisions of electrons with neutral particles in these discharges is fundamentally different: in DCD, electrons only lose energy as a result of collisions, and in RFCD the electrons can gain an essential energy of thermal motion only in the presence of collisions;

– a new external discharge parameter appears in the RF discharge – frequency of the RF field ω, depending on which the discharge properties change significantly for a fixed amplitude of the RF voltage;

– in the DCD, there are always two electrodes (cathode and anode) that functionally differ significantly, and in the RFD both electrodes are similar in their role to a divided cathode; in addition, the RFD can be 'single-electrode', when the surrounding electrode space is the 'earth' as the second electrode;

– in contrast to the DCD, in the RFCD the near-electrode layers of the space charge can not only change their parameters during the RF field period, but also undergo a complete collapse for a short time;

– if the cathode drop U_k in the DCD can not exceed the voltage U applied to the discharge gap, then the the instantaneous value of the voltage $V(t)$ in the near-electrode region of the RFCD can significantly exceed the amplitude of the applied RF voltage $V_\sim$ (in the asymmetric RFCD it can be $V(t) \leq 2V_\sim$);

– in contrast to the DCD, resonant oscillatory processes at the frequency of the RF field, its harmonics and other frequencies are possible in the RFCD;

– the existence of a stationary DCD is due to the important role of electron emission processes at the cathode, while the RFD is realized in the form of two of its varieties: a γ-discharge with mandatory emission of electrons from the surface of the electrodes and an α-discharge in which electron emission is not important for its maintenance;

– the DCD is created only by a potential electric field, and the RFD is created by both potential and vortex fields.

Due to the ease of excitation, the extraordinary variety of structural configurations of RF systems, the wide range of plasma parameters, the possibility of creating a plasma of chemically aggressive media with the help of external electrodes and the possibility of using a large number of factors to control the plasma parameters, the RFCD has found much more practical application than the DCD.

1.3. Electrical breakdown of gas in the RF field

Electrical breakdown is usually called the transformation of a non-conductive substance, including gas, into a conductor as a result of applying a sufficiently strong electric field to it.

In the case of a single mechanism for the escape of electrons from the discharge gap – diffusion, the equation of the balance of the number of electrons takes the form:

$$\frac{dn_e}{dt} = -D_e\Delta^2 n_e + v_i n_e = 0, \tag{1.13}$$

where D_e is the diffusion coefficient of electrons, n_e is the electron concentration, v_i is the ionization frequency by electron impact. Hence we obtain the breakdown condition:

$$D_e\frac{n_e}{L^2} + v_i n_e = 0 \quad \text{and further} \quad \frac{D_e}{L^2} \approx v_i,$$

where L is the diffusion length of the discharge gap.

As noted by S. Brown [157], there are significant differences in the nature of the breakdown in a gas initiated by the application of a constant and RF electric fields. In the RF discharge, the primary ionization caused by the vibrational motion of electrons is the only process that supplies electrons. Therefore, this type of discharge is the simplest by the mechanism. Gas breakdown under the action of a constant field is a more complicated phenomenon than an RF breakdown, since the electrons are continuously carried away from the discharge gap by an applied field. Because of this, additional physical processes are needed that compensate for the mentioned electron losses. In this sense, breakdown in a constant field is a more general case.

We recall first the well-known Townsend criterion for the breakdown, derived for the so-called avalanche discharges of the DCGD type and giving basic ideas about the laws of the phenomenon of gas breakdown. Then we note the features that appear in the case of RF breakdown.

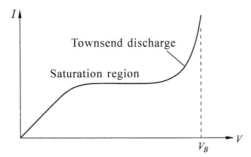

Fig. 1.3. Pre-breakdown I–V characteristic [158].

A typical current–voltage characteristic of the prebreakdown phenomena in the discharge gap is shown in Fig. 1.3.

As can be seen in Fig. 1.3, as the voltage V increases, the current increases with increasing speed and a voltage V_B is reached, at which the current value is limited only by the resistance of the external electrical circuit. The value V_B is called the breakdown voltage, which depends on the pressure and kind of gas, the distance between the electrodes d, the material, and the shape of the electrodes. The primary element of the breakdown process is the electron avalanche, which develops in a gas under the action of a field.

As is known [4], the Townsend breakdown criterion has the form:

$$\gamma \exp(\alpha d) = 1, \tag{1.14}$$

where d is the interelectrode distance, α and γ are the Townsend coefficients of bulk and surface ionization, respectively.

Using expression (1.14), one can find the dependence of the breakdown voltage V_B on the physical conditions in the discharge, i.e., the Paschen law [4]:

$$V_B = f(pd). \tag{1.15}$$

The experimental Paschen curves are shown in Fig. 1.4.

It should be emphasized that the Paschen law is valid only for discharges with moderate pressures p and discharge gaps d, when electronic avalanches are the determining element of the physical mechanism of discharge of a glow type. This law does not work at too high values of p and d, when the discharge becomes an arc discharge, in which other physical processes of discharge maintenance prevail – thermal ionization and volumetric recombination.

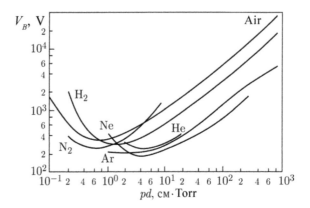

Fig. 1.4. The $V_B(pd)$ experimental dependence [158].

Let us now consider the specific features of the RF breakdown of the gas. In this case, the character of the breakdown depends essentially on the frequency of the field ω.

The breakdown voltage V_B will be large at both high and low frequencies of the RF field:

1) at high frequencies ($\omega \gg v_{en}$) the energy deposition of the RF field $P_\sim$ into the gas is small:

$$P_\sim = jE_\sim = \sigma E_\sim^2 = \frac{e^2 n_e v_{en} E_\sim^2}{m(\omega^2 + v_{en}^2)},$$

where $\sigma = \dfrac{e^2 n_e v_{en}}{m(\omega^2 + v_{en}^2)}$ is the conductivity of an ionized gas, $E_\sim$ is the RF field strength, v_{en} is the collision frequency of electrons with gas atoms, so V_B must be large;

2) at low frequencies ($\omega \ll v_{en}$), the energy set by electrons between two collisions is small, and high voltages V_B are needed to develop the breakdown.

We estimate the optimal conditions for the effective energy deposition of the RF field in the discharge. For this it is necessary to find the optimal relationship between the values of ω and v_{en} from the condition:

$$\frac{dP_\sim}{dv_{en}} = 0.$$

Hence the condition for the optimal energy input has the form:

$$\omega = V_{en}. \tag{1.16}$$

On the Paschen curves for the RF discharge, the type shown in Fig. 1.4, a minimum of V_B will be observed where condition (1.16) is satisfied. When the family of curves $V_B(pd)$ is obtained for different frequencies ω, the position of the minimum will change.

Due to frequency dependence of V_B, an interesting phenomenon of the heavy dependence of electron energy accumulation on the parameters of RF discharge, such as field frequency ω and gas pressure p, is present. Thus, it has been established that V_B depends on the RF field period $T = 2\pi/\omega$ and characteristic time $\tau = 1/\delta v_{en}$ ratio, where $\delta = 2m/M$, m and M – electron and atom masses, respectively, τ – is the time at which the electrons lose entirely their excessive energy gained in the electric field.

Let us consider the cases of low and high frequencies of the field (Fig. 1.5).

a) Low frequencies ($T > \tau$). In this case, the change in the average electron energy $\bar{\varepsilon}_e$ follows the change in $E^2_{\sim}$, as shown in Fig. 1.5 *a*. The development of the discharge should occur in a time shorter than 1/4 period of the RF field. In this case, $\bar{\varepsilon}_e \approx \mathrm{const}$.

b) High frequencies ($\tau > T$). Here it is possible to collect energy by electrons over several periods of the RF field. This is the hallmark of RF breakdown, in which there can be $V_B \le V_i$, where V_i is the

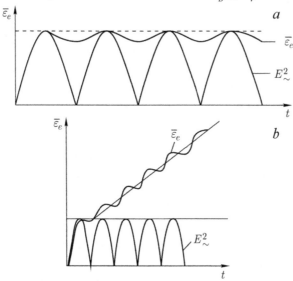

Fig. 1.5. The nature of the process of energy collection by electrons in the RF field. *a*) low frequency ($T > \tau$). *b*) high frequency ($\tau > T$).

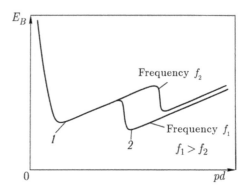

Fig. 1.6. The Paschen curve for RF breakdown in a gas with two minima [158].

ionization potential of the gas atom. Here $\bar{\varepsilon}_e \neq$ const, as is seen in Fig. 1.5 *b*.

We note the specific form of the Paschen curve for RF breakdown, in comparison with the analogous curve for breakdown in a gas in a constant field. The experimental Paschen curve for RF breakdown is shown in Fig. 1.6. From this it is clear that the curve $V_B(pd)$ at a fixed frequency ω has two minima.

The first minimum is similar in nature to the minimum that occurs when a breakdown takes place at a constant current. The second minimum arises when the amplitude of the oscillations of the electrons A becomes of the order of or less than the interelectrode distance d, i.e. $A \leq d$, which leads to a significant deceleration of the escape of electrons from the discharge. Further rise in the $V_B(pd)$ curve after the second minimum is due to the same reason as after the first minimum – a decrease in the energy set by electrons between collisions with atoms.

In order to obtain in-depth ideas about the physical phenomenon of RF breakdown in gas and the parameters of gas-discharge processes, a more detailed study of the physical meaning of the Paschen curves of ignition of the $V_B(pd)$ discharge was carried out in the last period.

An interesting feature of the behaviour of the ignition curve was discovered in Ref. [23]: there exists a region of an ambiguous dependence of the ignition voltage of the discharge V_B on the gas pressure to the left of the minimum of the RF breakdown curves, as is clear from the later results of [178] in Fig. 1.7.

As recommended in [178], the RF curve of the gas breakdown consists of four branches differing in the nature of the processes participating in the breakdown: 1) multipactor, 2) Paschen, 3)

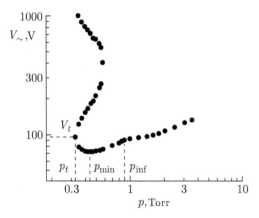

Fig. 1.7. The RF breakdown curve in hydrogen in the discharge gap with interelectrode distance d = 2 cm [178].

diffusion-drift and 4) emission-free. These branches can be designated on the breakdown curve, using for orientation the characteristic points of the latter with the coordinates (p_t, V_t), (p_{min}, V_{min}), (p_{inf}, V_{inf}) in Fig. 1.7.

Figure 1.7 shows that the point (p_t, V_t) is the turning point on the curve $V_B(pd)$, for which it is established that the amplitude of the displacement of electrons is equal to half the distance between the electrodes, which leads to increased losses of electrons on the electrodes [179]; the point (p_{min}, V_{min}) is the point where there is a minimum of the curve $V_B(pd)$ from the low-pressure side; the point (p_{inf}, V_{inf}) is the point at which the inflection on the curve $V_B(pd)$ is observed after its minimum on the side of small p, and at $p > p_{inf}$ it is established that the curve $V_B(pd)$ depends little on the material of the electrodes [180].

As shown in [180], the values of V_{inf} and p_{inf} can be found using the known values of V_{min} and p_{min}, according to the relations

$$\frac{V_{inf}}{V_{min}} = \frac{e}{2}, \quad \frac{e}{2} < \frac{p_{inf}}{p_{min}} < e,$$

where e is the base of the natural logarithms.

Let us consider the branches of the RF breakdown curve mentioned earlier.

The branch to the right of the inflection point $(p > p_{inf})$ is characterized by a weak dependence on the material of the electrodes, therefore we can assume that for $p > p_{inf}$, the emission of electrons

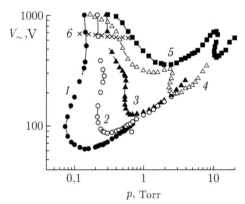

Fig. 1.8. The curves of RF breakdown in air: 1 – d = 29 mm; 2 – d = 20 mm; 3 – d = 14 mm; 4 – d = 9 mm; 5 – d = 6.5 mm and 6 – breakdown curve for RF dark discharge at d = 6.5 mm [178].

from the surface of the electrodes does not participate in the RF breakdown of the gas (the emission-free branch).

The so-called diffusion-drift branch of the breakdown curve exists in the pressure interval between the turning point p_t and the inflection point p_{inf}. In this case, the emission of electrons from the surface of the electrodes participates in the breakdown, along with the ionization of the gas atoms by electron impact, electron drift in the RF field, and electron losses on the electrode surfaces and the walls of the discharge chamber. The diffusion-drift branch is most pronounced for sufficiently long discharge gaps (d > 1 cm). At smaller discharge gaps (d < 1 cm), it is less pronounced, but a second minimum can be observed (Fig. 1.8). This second minimum is at pressures less than the pressures corresponding to the minimum of the diffusion-drift branch, and it is sometimes called the main (Paschen) minimum [158]. We will also call this branch of the breakdown curve in the form of a trial variant as *Paschenov's*.

At a sufficiently low gas pressure, the voltage of the RF breakdown depends only slightly on the gas pressure (Fig. 1.8). Perhaps, there is a transition to the regime of resonant RF breakdown controlled by electron emission [2]. This branch of the breakdown curve is often called *multipactor*. At low pressures, electrons move in the discharge chamber, colliding with the electrodes much more often than with neutral gas particles. Electrons can multiply themselves, if only they oscillate between the electrodes in resonance with the RF field and hit strongly enough electrodes to knock out secondary electrons.

Another circumstance is noted in [178]: a number of authors [2, 158, 167] assert that the condition $\omega \approx v_{en}$ is satisfied at the minimum of the Paschen branch. Meanwhile, the data of [178] presented in Fig. 1.8, show that under the conditions of the aforementioned minimum, values of $\omega = 8.5 \cdot 10^7$ s^{-1} and $v_{en} = 9.8 \cdot 10^9$ s^{-1} were obtained, i.e. $\omega \ll v_{en}$.

Studies have shown [178, 181] that the study of RF breakdown in a gas can yield important information. Let us dwell only on two examples: 1) determination of the electron drift velocity v_{dr} over a wide range of values of the reduced electric field strength E/p; 2) determination of the RF voltage $V_{\sim}$ at the active electrode in complex gas-discharge process units.

Let us consider a technique for measuring the drift velocity of electrons in an electric field and the possibility of using it to estimate the RF voltage at the active RFCD electrode in an installation.

To determine the drift velocity of the electrons v_{dr}, we use the coordinates of the turning point (p_t, V_t) on the RF breakdown curve (Fig. 1.7). Since this point is usually well expressed, the process of measurement does not cause any difficulties.

Let us consider the motion of electrons in a homogeneous RF electric field. The drift velocity of the electrons in the RF field in the case $v_{en} \gg \omega$ is described by expression

$$v(t) = \frac{eE_{\sim}}{mv_{en}}\cos \omega t. \tag{1.17}$$

The amplitude of the drift velocity

$$v_{dr} = \frac{eE_{\sim}}{mv_{en}}$$

is the maximum instantaneous electron velocity corresponding to the amplitude value of the RF field strength. Integrating expression (1.17) with respect to time, we obtain the amplitude A of the electron displacement in the RF field:

$$A = \frac{eE_{\sim}}{mv_{en}\omega} = \frac{v_{dr}}{\omega}.$$

At the turning point of the RF ignition curve with coordinates $p = p_t$ and $V_{\sim} = V_{\sim t}$, the electron displacement amplitude is $A = d/2$, and therefore for the drift velocity we get a simple expression:

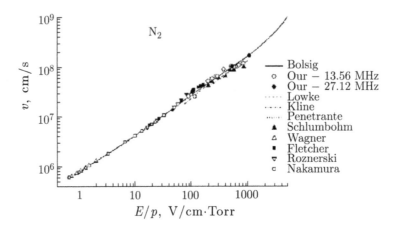

Fig. 1.9. Dependence of the drift velocity v_{dr} of electrons in nitrogen on $E_\sim/p$ [181].

$$v_{dr} = d\pi f, \tag{1.18}$$

where $f = \omega/2\pi$. At a fixed value of the frequency of the RF field f and the distance d between the electrodes, the drift velocity of the electrons at the turning point is constant, independent of the gas composition.

The results of measurements of v_{dr} obtained in [181] by the described method, together with the calculated and experimental data of other authors, are shown in Fig. 1.9. As can be seen, the results obtained are in good agreement with the literature data.

Thus, the measured RF ignition curve makes it possible to determine one value of the drift velocity. To determine a number of values of the electron drift velocity for different $E_\sim/p$, it is necessary to measure a family of RF ignition curves for different interelectrode gaps and / or frequencies of the RF electric field.

The measured coordinates of the turning point allow us to determine the ratio of the electric field strength to the gas pressure $E_\sim/p$, corresponding to the measured drift velocity of the electrons.

Let us consider the application of the method described above by the authors of Ref. [181] to estimate the magnitude of the RF voltage $V_\sim$ on an active electrode under conditions of a complex process unit, where access to the electrode is difficult.

Let the distance between the flat electrodes in the process unit be d, and the frequency of the RF generator f. Gas is introduced into the discharge chamber, preferably nitrogen, for which the electron drift velocity is known with good accuracy. Further, the ignition curve of

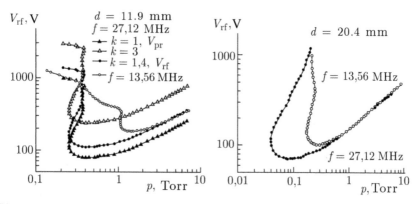

Fig. 1.10. a) RF ignition curves in nitrogen for $d = 1.19$ cm and $f = 27.12$ MHz: measured with the RF probe $V_{pr}(p)$ ($k = 1$) and refined by circuit correction factors ($k = 3$) and ($k = 1.4$), as well as the RF ignition curve for $f = 13.56$ MHz; b) RF ignition curves in nitrogen for $d = 2.04$ cm, $f = 27.12$ MHz, $f = 13.56$ MHz [181].

the RF discharge $V_\sim(p)$ is measured. Since this measurement is carried out using a standard RF probe connected to a connector outside the discharge chamber and giving approximate measurements, then in reality a certain dependence $V_{pr}(p)$ is obtained. In the region to the left of the minimum of the ignition curve, we obtain the turning point with the coordinates $p = p_t$ and $V_{pr} = V_{t.pr}$. At this point, the electron drift velocity according to (1.18) can be easily determined using the known values of d and f. Let us consider a specific example for the RF ignition curve for nitrogen ($d = 1.19$ cm, $f = 27.12$ MHz) in Fig. 1.10 [181]. For it, at the turning point, the electron drift velocity is equal to $v_{dr} = d\pi f = 1.01 \cdot 10^8$ cm/s.

Then, from Fig. 1.9 it is determined that in order for electrons in nitrogen to have such a drift velocity, one must have $E/p = 530$ V/ (cm $\cdot$ Torr). The coordinates of the turning point for this ignition curve are equal to $p_t = 0.25$ Torr and $V_{t.pr} = 113$ V. Since $E = V_\sim/d$ ($V_\sim$ is the amplitude of the RF voltage on the active electrode), then from the known value of E/p we have:

$$\frac{E}{p_t} = \frac{V_\sim}{p_t d} = 530\,V\,/\,(\text{cm}\cdot\text{Torr}),$$

$$V_\sim = 530 p_t d = 530\cdot0.25\cdot1.19 = 158.3\ \text{V}.$$

Then the real RF voltage $V_\sim$ on the active electrode is connected with RF voltage V_{pr}, measured with an RF probe outside discharge chamber, by the relation

$$V_{\sim} = 1.4 V_{pr}.$$

Additional details of this diagnostic technique are given in [181].

1.4. Power transmission processes in the RFCD network

It should be emphasized that it is incorrect to study the physical properties of the RF discharge in isolation from its full electrical circuit. The discharge occurring in the interelectrode space in the general case can have not only a complex impedance with active and reactive components, but also its own internal electromotive force (EMF), being only a part of the whole electrical discharge circuit. The output voltage of the RF generator falls on all elements of the circuit in proportion to the magnitude of their impedances.

The electric circuit of the RFCD includes: an RF generator, a matching device, a meter of the RF power input to the discharge, a blocking capacitance and a discharge gap. The blocking capacitance C_{bl} interrupts the discharge circuit in direct current before the discharge gap. The C_{bl} value should not be small so that the greater part of the RF voltage is not uselessly falling on the impedance of the capacitance. At the same time, it must be taken into account that in the non-contact method for diagnosing RFCD described below it is necessary to measure the constant component of the voltage at this capacitance, which is the smaller, the larger C_{bl}. Therefore, we must take its optimal value.

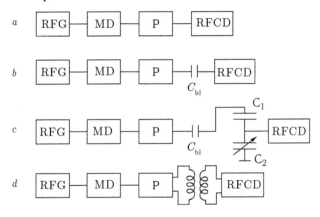

Fig. 1.11. Methods for bringing RF power from the RF generator to the RFCD. RFG – RF generator, MD – matching device, $P_{\sim}$ – RF power meter. *a*) discharge circuit without blocking capacitance C_{bl}, *b*) discharge circuit only with a blocking capacitance C_{bl}, *c*) discharge circuit with C_{bl} and a capacitive RF voltage divider, *d*) a transformer method of supplying RF power to the discharge.

The RF generator, depending on the technical requirements for the experimental conditions, can operate in two modes: 1) a voltage generator or 2) a current generator. In the first case, the internal resistance of the oscillator R_{int} is small compared to the total resistance of the entire discharge circuit R_Σ, and in the second it will be $R_{int} \gg R_\Sigma$.

There are various ways of bringing the RF power to the discharge, shown schematically in Fig. 1.11.

The method of transformer coupling in Fig. 1.11 *d* is used in case of needing breaking the galvanic connection between the output stage of the RF generator and RFCD.

The passage of RF power in the full electric circuit of RFCD is accompanied by its dissipation on practically all elements of the radio engineering tract. The energy of the RF field partially flows through the frequency-dependent parasitic capacitances, dissipates in the MD, is released at the blocking capacitance C_{bl}, is absorbed by the dielectric walls in the presence of external electrodes, is dissipated as a result of edge effects at the ends of the discharge gap, and finally absorbed by various mechanisms directly in the RFCD (the ionization and excitation of a neutral gas, the acceleration of ions and emitted electrons in the NESCL, stochastic heating of plasma electrons at the oscillating boundary of the NESCL, excitation of various instabilities in the plasma with the generation of intense microwave fields and others). It should be noted that up to 50% of the applied RF power can be consumed at low gas pressures to accelerate ions in the NESCL. In the literature there are judgments that these are net losses that significantly reduce the effectiveness of RFCD as a plasma generator. It seems that this interpretation is very simplified, since in this case the intensity of electron emission increases due to intensive bombardment of the electrodes by ions and the density and energy of the electron beams supporting the discharge increase.

The parameters of the edge effects depend on all the RFCD parameters, and the edge effects themselves can initiate secondary effects, for example, the generation of wall electron beams that returns part of the energy leaving the RF field back to the discharge.

Power transmission processes depend significantly on the configuration of the discharge gap (symmetrical and asymmetric RFCD). The discharge asymmetry can be of an electrical nature (one electrode is grounded) or geometric nature (the electrode areas differ significantly).

In the literature [6, 174] it is shown that a symmetrical RFCD is a linear radio engineering system in which there are no non-linear effects, for example, harmonics of the discharge current. In the asymmetric discharge, non-linear effects, on the other hand, are strongly pronounced, in particular, a strong anharmonicity of the discharge current is observed, and a quasi-electrostatic EMF arises together with a constant component of the discharge current.

We note that in the simpler case of a symmetric RFCD, energy transfer processes can be complicated. Thus, for a relatively large interelectrode distance d, a symmetrical RFCD can consist of two autonomous single-electrode discharges. Moreover, as d decreases, the process of energy transfer can change significantly: two single-electrode discharges are converted into one two-electrode discharge and the conditions regarding the onset of beam-plasma instabilities in the discharge can radically change.

The energy transfer processes, naturally, are reflected in the I–V characteristic of the RFCD, since at each of its points the power absorbed in the discharge $P = I_{\sim} V_{\sim} \cos \varphi$ varies, where φ is the phase shift between the current $I_{\sim}$ and the RF voltage $V_{\sim}$.

Attention is drawn to the fact that the primary energy of the RF field from the RF generator in RFCD repeatedly turns into other types of energy. So, the RF field, getting into the near-electrode layer of the space charge, is rectified in it because of its non-linear current–voltage characteristics, analogous to that for an ordinary diode. The quasi-stationary field that arises in the NESCL accelerates the electrons emitted by the electrode, converting them into near-electrode electron beams with high kinetic energy. In addition to active ionization and excitation of a neutral gas, these beams can excite beam-plasma instabilities, which in turn generate intense electromagnetic fields in the microwave range. The latter additionally and significantly heat plasma electrons, which leads to an intensification of all the elementary processes in the plasma.

From the foregoing, we can conclude that the processes of energy transfer in the electric circuit of RFCD are very complex. In an equivalent circuit, RFCD should, as far as possible, reflect the most important elements in which the basic processes of energy dissipation take place. One of the simplest versions of the equivalent RFCD scheme is shown in Fig. 1.12.

Of course, due to the great difficulties in formalizing all the most important elements of the physical mechanism of RFCD in the form of concentrated parameters of an equivalent circuit, it is

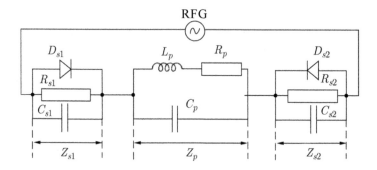

Fig. 1.12. Equivalent circuitry of RFCD. C_{s1}, C_{s2}, R_{s1}, R_{s2}, D_{s1}, D_{s2} – capacitances, active resistances and conditional diodes corresponding to the NESCL; C_p, R_p, L_p – capacity, active resistance and plasma inductance.

not necessary to exaggerate the importance of such an approach to studying the discharge, but it is also useful.

1.5. Current–voltage characteristic of the RFCD

The most general, macroscopic characteristic of a RF discharge is its current–voltage characteristic (CVC). At the same time, the CVC contains a lot of information about the discharge both about the physical phenomenon as a whole, and about the parameters of the NESCL and the RFCD plasma in the cases of certain regimes of the latter, considered below.

We note a number of informative moments obtained in the analysis of experimental I–V characteristics:

1) it determines the range of variation of the RF voltage $V_\sim$ and current $I_\sim$ under the specific experimental conditions, for example, in a discharge tube with the given parameters;

2) the qualitative course of the CVC reflects the change in physical conditions in the discharge, the transitions between its various regimes;

3) regions of instability of the discharge parameters are observed, for example, a jump in the parameters during the transition from the α-discharge to the γ-discharge;

4) with an additional measurement of the phase shift φ between the current $I_\sim$ and the voltage $V_\sim$, one can measure the power enclosed in the discharge $P_\sim = I_\sim V_\sim \cos\varphi$ at each point of the I–V characteristic;

5) the most important point in the research is the choice of working points on the I–V characteristic, for example, when studying α- or γ-discharges, normal or abnormal conditions of RFCD;

6) it is possible to judge the correctness of the model representations by comparing the experimental CVC and the CVC calculated for a specific physical model of the discharge;

7) a CVC-method for diagnosing RFCD based on obtaining a family of CVCs with different interelectrode distances d and using a justified equivalent electric discharge circuit, which allows determining a number of parameters of near-electrode layers and gas-discharge plasma, is developed.

It should be noted that in the literature, the question of correctly obtaining the CVC characteristics and its interpretation is often not enough discussed. In this connection, one should mention CVC for such cases as single-electrode RFCD, symmetrical discharge with variable interelectrode distance d, discharge with one grounded electrode, for which the I–V characteristic in a circuit of active and grounded electrodes can vary greatly, especially if the values of d and $V_\sim$ also vary greatly.

A block diagram of the experimental setup for studying the I–V characteristic curves of the RFCD is shown in Fig. 1.13.

The alternating current $I_\sim$ in the electric circuit of the RFCD is measured by various methods: 1) using the Rogowski belt and the electric circuit shown in Fig. 1.13, 2) with thermomillammeters (type T-217), 3) by means of an oscilloscope recording of the voltage at

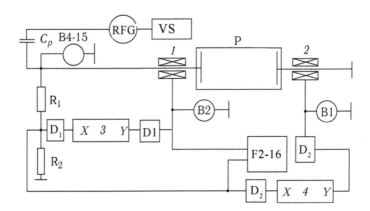

Fig. 1.13. Block diagram of the experimental setup for studying the I–V characteristic curves of the RFCD. *1, 2* – Rogowski's belts; *3, 4* – two-coordinate recorders; P – discharge tube; B1, B2, B4-15 – RF voltmeters; D_1, D_2 – detectors; Φ2-16 – phase meter; R_1, R_2 – the RF voltage divider; VS – voltage source.

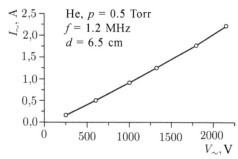

Fig. 1.14. CVC of the low-pressure RFCD.

the low-inductance resistance in the discharge circuit, 4) using a computational oscilloscope (type TDS), to which the signals from the Rogowski belt and RF voltmeter outputs, digitized by the ADC, are applied.

Let us consider the CVC of the RFD for the different physical conditions, and discuss their interpretation.

An example of the current–voltage characteristic obtained with a low gas pressure is shown in Fig. 1.14. It can be seen that the discharge current $I_\sim$ increases monotonically with the voltage $V_\sim$. There is no structuring of the I–V characteristic, as a result of which it is not possible to identify any discharge mode.

A great help in interpreting the CVC for the RFCD is due to a well-studied I–V characteristic for a wide range of DC-current discharge currents, depicted in Fig. 1.15 [151]. This I–V characteristic has a clearly expressed structure, which makes it possible to distinguish different types and regimes of gas discharges. First of all, interest is in area *3* (normal glow discharge), *4* (anomalous glow discharge), and *5* (transition to arc discharge) having their RF counterparts. Carrying out the corresponding analogies in studying the physical mechanism of RFCD should be carried out fairly correctly, taking into account that in this case the total discharge current consists of the conduction current and the bias current. The contribution of these components in the two types of glow RFCD (*α*-discharge and *γ*-discharge) will be different.

For a long time, the attention of researchers was attracted by the so-called 'normal' glow discharge regime (section 3 of the current–voltage characteristic in Fig. 1.15). It is distinguished by the constancy of the values of the discharge-supporting voltage U_n, the current density j_n, and the thickness of the NESCL d_{sn} in a certain range of discharge currents under given experimental conditions.

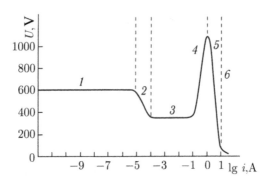

Fig. 1.15. CVC of various types of DC discharge in neon with copper electrodes [151]. $p = 1$ Torr, $d = 50$ cm. *1* – Townsend discharge; *2* – transition to a glow discharge, *3* – normal glow discharge; *4* – anomalous glow discharge; *5* – transition to the arc; *6* – arc discharge.

The effect of normal mode in DCGD and RFCD is considered in detail in the monograph [6]. Let us cite a number of basic points.

This effect in DCGD has been known for a long time. The discharge occupies such an area on the cathode that the current density at the cathode has a certain 'normal' value, depending on the kind and pressure of the gas and electrode material. It was established earlier [165] that this effect is related to the existence of a minimum on the CVC of the cathode layer $V_c(j)$. When the cathode is not completely filled with current, a state corresponding to the minimum ('normal') cathode drop of the potential $V_c \approx V_{min} = V_n$, which corresponds to the normal current density j_n, is realized. The question of the physical reason for the existence of the voltage V_n has been elucidated relatively recently [166, 167]. It was found that the reason is the instability of the discharge with a cathode drop $V_c > V_{min}$. On the left (falling) branch of the current–voltage characteristic $V_c(j)$, the state inside the cathode layer is unstable, which was understandable for a long time. On the right (growing) branch of the CVC, where the states are stable, the state at the edge of the layer is unstable at the boundary between the current and non-current zones at the cathode. But this became clear much later [166]. If the cathode is filled with current in its entirety, there is no boundary between the current and non-current zones on the cathode. Since the state inside the layer on the right branch is stable, it is realized in practice, corresponding to the anomalous glow discharge: $V_c > V_n$, $j > j_n$.

A situation similar to the situation described above is also carried out in the α-type of RFCD. The reason for the effect of the normal current density is the existence of a minimum on the CVC of the whole α-discharge as a whole [66]. According to experimental studies, in the RF discharge the CVC of the plasma of the positive column $V_{PC}(j)$ has a decreasing nature, and the CVC of the near-electrode layers, where the bias current flows, is increasing. Consequently, the voltage at the electrodes $V_\sim$, which consists of the voltages on the layers and the plasma, must have a minimum, depending on j. In [6] an expression was obtained for the normal current density corresponding to this minimum:

$$j = j_{na} = (\omega Cpd\sqrt{m} / 4\pi d_\alpha)^{1/(m+1)},$$

where C and $m > 0$ are certain constants. Hence it is clear that j_{na} is the larger the higher the pressure p, the longer the discharge gap d and the higher the frequency of the field ω. This explains why the α-discharge has an upper bound on p and d. With increasing pd, j_{na}, field strength and voltage drop in the NESCL increase. At a certain critical value of the value of $(pd)_{cr}$, the voltage on the layer of the normal α-discharge becomes so large that the NESCL breaks and the α-discharge passes into the γ-discharge, that is, when $(pd) > (pd)_{cr}$ α-discharge can not exist. The quoted work also shows that in the α-discharge, with an appropriate decrease in the frequency of the field and pressure, the effect of the normal current density disappears.

As was experimentally established [6], in the range of middle pressures the normal mode takes place both in the γ-type and α-type

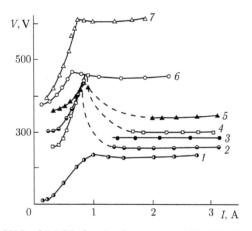

Fig. 1.16. The CVC of RFCD for the frequency of 13.56 MHz with the insulation of the electrodes with Teflon: *1* – helium (p = 30 Torr, d = 0.9 cm), *2* – air (30 Torr, 0.9 cm), *3* – air (30 Torr, 3 cm), *4* – air (30 Torr, 0.9 cm), *5* – CO_2 (15 Torr, 3 cm), *6* – air (7.5 Torr, 1 cm), *7* – air (7.5 Torr, 1 cm) [6].

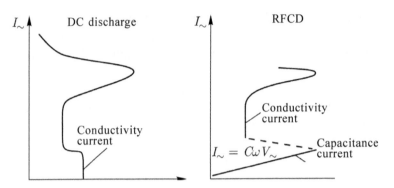

Fig. 1.17. CVC of DCGD and RFCD at medium pressure.

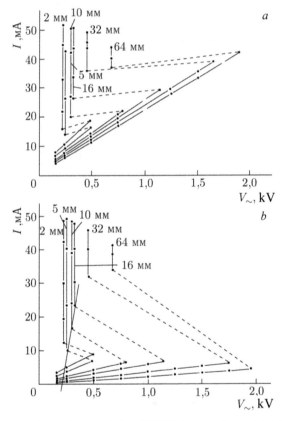

Fig. 1.18. Current–voltage characteristic of RFCD (active electrode) (*a*) and (grounded electrode) (*b*) for various interelectrode distances. Air, 30 Torr, 1.2 MHz.

RFCD, although in the latter case it occupies a small current range (Fig. 1.16).

At slightly elevated pressures, the I–V characteristic acquires the form containing elements of the I–V characteristic of the DCGD.

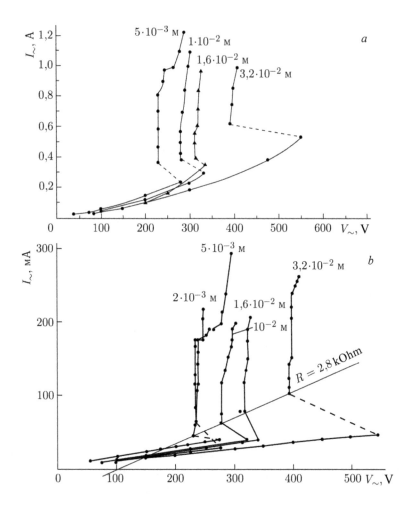

Fig. 1.19. CVC of RFCD (active electrode) (*a*) and (grounded electrode) (*b*) for various interelectrode distances. Helium, 50 Torr, 1.2 MHz.

This can be seen in Fig. 1.17, which qualitatively compares the I–V characteristic of these discharges.

The CVC of middle-pressure RFCDin air and helium obtained in the electrical circuit of a grounded electrode (GE) are shown in Figs. 1.18 and 1.19 respectively.

Interpretation of the structureless I–V characteristic of the low-pressure RFCD in Fig. 1.14 becomes more understandable after considering the evolution of the type of CVC of the RFCD with the pressure decrease shown in Fig. 1.20.

On the basis of the material given earlier, let us analyze the physical meaning of the experimental CVC for the RFCD.

The parts of the current–voltage characteristic of the DCGD, shown in Fig. 1.15, are well known: *1* – Townsend discharge, *3* – normal DCGD mode, *4* – anomalous DCGD mode, *5* – transition from glow discharge to arc discharge, *6* – arc discharge.

The comparison of the I–V characteristic of the DCD and the RFCD at a qualitative level shown in Fig. 1.17 displays the presence of common areas in their nature, namely, sections *2*, *3*, and *4*. At the same time, the sections with the minimum current values differ in their type of current–voltage characteristic: the Townsend discharge and the lower, flat part of the CVC of the RFCD, which varies linearly. At the end of the latter, a breakdown of the discharge gap occurs, and the RFCD occurs at the beginning of the previously mentioned section *2* – the RF analogue of the section of the normal DC discharge mode.

The lower sloping branch of the current–voltage characteristic of the RFCD before the breakdown is the bias current flowing as a result of applying a harmonic RF voltage $V_\sim = V_{\sim 0} e^{i\omega t}$ to the capacitance of the discharge gap C. The instantaneous value of the charge in this capacitance is $Q = CV_\sim = CV_{\sim 0} e^{i\omega t}$. Hence the bias current is

$$I_\sim = \frac{dQ}{dt} = C\omega V_\sim,$$

i.e., it varies linearly, as on the lower branch of the current–voltage characteristic of the RFCD.

The described situation takes place if the physical conditions are such that the RFCD occurs immediately in a relatively high-current form (γ-mode). However, physical conditions are possible, when a low-current discharge form, the α-mode, occurs during the passage of the lower branch of the current–voltage characteristic. Then, beginning with a certain value of $V_\sim$, the current on the lower branch increases slightly, according to a non-linear law, until the jumplike transition to the γ-mode.

Consideration of the evolution of the I–V characteristic curves with a monotonic decrease in the gas pressure in Fig. 1.20 shows that at the same time the current interval of the normal mode is reduced until it disappears. A further decrease in pressure leads to a transformation of the form of the I–V characteristic, which looks like a smooth transition of the α-branch to the anomalous mode of the γ-branch of the CVC for the RFCD.

Let us now analyze in more detail the physical meaning of the CVC of the RFCD of medium pressure in air and helium with a clearly expressed structure (Figs. 1.18, 1.19).

First of all, it follows from general considerations that the true characteristic of RFCD is its I–V characteristic obtained in the circuit of a grounded electrode (Figs. 1.18 *b*, 1.19 *b*). In fact, the current is closed at the grounded electrode (GE) from the active electrode (AE), which has passed through the entire discharge gap and carried out the energy deposition in the whole set of physical processes supporting the discharge. The nature of the experimental VACs under consideration confirms this. First, the current $I_{\sim AE}$ in the AE circuit for many gases is much greater than that of $I_{\sim GE}$ in the GE circuit (Fig. 1.18, 1.19). In this case, most of the current $I_{\sim AE}$ is closed directly to the 'ground' surrounding the discharge through parasitic capacitances. Possible deviations from this are explained below. Secondly, we pay attention that in the CVC family for AEs with different interelectrode distances d (Figs. 1.18 *a*, 1.19 *a*), the jumplike transition from the α- to the γ-branch occurs under significantly different physical conditions in the discharge, determined by the current $I_{\sim AE}$. Meanwhile, in the analogous family of I–V characteristics for the GE (Figs. 1.18 *b*, 1.19 *b*), the $\alpha \to \gamma$-jump occurs always at the same value of $I_{\sim GE}$, which provides the physical conditions for the transition between these branches. This

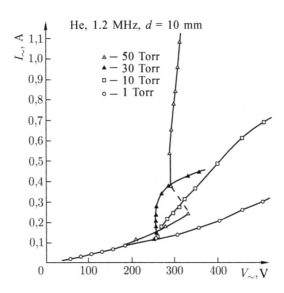

Fig. 1.20. Evolution of the current–voltage characteristics of the RF with decreasing gas pressure.

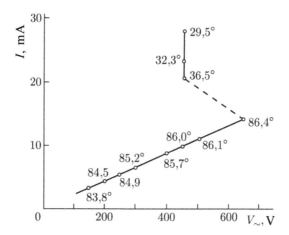

Fig. 1.21. Dependence of the phase shift between the RF current and the voltage on the position on the CVC of the RFCD (grounded electrode). Air, 10 Torr, 1.2 MHz, $d = 52$ mm.

unequivocally indicates that only the current–voltage characteristic in the GE circuits correctly reflects the state of the RFCD under consideration.

So, the CVC of the middle pressure consists of two branches – lower and upper. An examination of the jump point of the voltage from the lower branch to the upper branch shows that at this point the voltage $V_{\sim}$ is close to the breakdown voltage V_{B} on the Paschen curve $V_{\text{B}}(pd)$ for a given gas. From here we can assume, in the first approximation, that the lower branch represents a pure bias current, and after the breakdown RFCD emerges immediately as an upper branch. The latter is analogous to the section of the 'normal' DCGD mode (Fig. 1.17), where there is only an active conduction current. It has been experimentally established that after a jump in the I–V characteristic, the phase shift between the current $I_{\sim}$ and the voltage $V_{\sim}$ decreases sharply (Fig. 1.21). This means that the contribution of the conduction current $I_{\sim a}$ to the total discharge current $I_{\sim}$ increases sharply in this case.

When obtaining the I–V characteristic curves of the type shown in Figs. 1.18, 1.19, it is established that in the vertical section of the current–voltage characteristic the investigated RF discharge behaves in all respects similarly to DCGD in the normal mode. In this case, the glow occupies only a part of the area of the discharge electrodes, the conduction current density $j_{\sim a}$ is close to the normal current density j_n in the DCGD, after the entire area of the electrodes is filled with luminescence the CVC path deviates from verticality (Fig. 1.19).

Experimental studies of DCGD and RFCD in the same discharge tube showed that it is possible to introduce for the RFCD the same parameters of the normal mode as for DCGD: 1) the normal near-electrode voltage drop $V_{\sim n}$, 2) the normal conductivity current density per electrode $j_{\sim n}$ and 3) the normal width of the near-electrode drop (the normal width of the NESCL) d_{ns}. Moreover, the quantities $V_{\sim n}$, $j_{\sim n}$, d_{ns} in the RFCD are close in magnitude to those in the normal DCGD mode.

We note a very significant dependence of the behaviour of the CVC of the RFCD on the kind of gas, which was observed in the study of these discharges in air and helium.

Comparing the I–V curves in air for the AE and GE circuits (Fig. 1.18), we find their difference in behaviour on the lower and upper branches: the currents of the lower branch in the AE chain are much larger than those in the GE circuit, and after the jump to the upper branch, the currents in both circuits practically compare. The latter means that after the jump the RFCD in the air becomes two-electrode: the current of the active electrode is almost completely closed to the grounded electrode. Note that the discharge is in the normal mode.

A different behaviour of the I–V characteristic curves is observed with a discharge in helium (Fig. 1.19). Here, too, the currents of the lower branches for the AE and GE circuits differ significantly, however, unlike the discharge in the air, and after the jump, the currents in the GE circuit remain much smaller than the currents in the AE circuit. Here, too, the RFCD switches to normal mode.

It seems that it is possible to explain such behaviour of RFCD in air and helium as follows. According to V.L. Granovsky [151], the parameters of the normal DCGD mode in air and helium are characterized by the following quantitative data: for air, $j_n/p^2 = 400$ $\mu A/(cm^2 \cdot Torr^2)$, for helium, $j_n/p^2 = 2.2$ $\mu A/(cm^2 \cdot Torr^2)$. Thus, the normal current density for air is almost 200 times greater than that for helium. In this regard, to ensure the normal mode of RFCD in the air, almost all the current in the AE circuit should be closed to the GE. In the case of RFCD in helium to ensure the normal discharge mode is sufficient to short-circuit only a small part of the current in the AE circuit to the grounded electrode. At the same time, the rest of the current in the AE circuit is still closed to 'ground' surrounding the discharge.

A complete analogy in the behaviour of the current–voltage characteristics of the RFCD and DCGD, which are in the normal mode, is observed when considering the families of the current–voltage characteristics in Figs. 1.18, 1.19. Indeed, with a monotonic decrease in the interelectrode distance d, the RF voltage $V_{\sim n}$, which maintains the normal mode, decreases. As shown below, the spatial structure of RFCD contains all the spatial regions of the DCGD: the region of near-electrode drop $V_{\sim ns}$ (cathode drop), the negative glow region, the Faraday dark space and the positive column of the

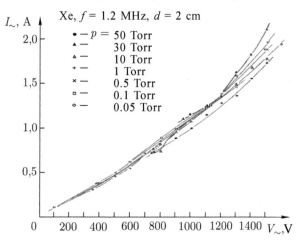

Fig. 1.22. CVC with the external electrodes, dielectric – quartz ($\varepsilon = 5$).

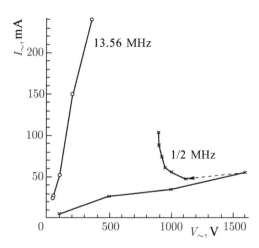

Fig. 1.23. CVC of RFCD with external electrodes, dielectric – TiBa ($\varepsilon = 2000$).

plasma. Thus, the RF voltage in the normal mode $V_{\sim n}$ is composed of the normal near-electrode drop $V_{\sim ns}$ and the sum of the drops on all other areas.

As can be seen from the families of the CVC curves of the RFCD (Figs. 1.18, 1.19), with a decrease in the interelectrode distance d (when different spatial regions of the discharge are removed), the voltage $V_{\sim n}$ decreases. Wherein the minimum value of $V_{\sim n}$ is observed when the boundary of the near-electrode region reaches $d = d_{ns}$. It was found experimentally that further decreasing d and maintaining the magnitude of the current $I_{\sim}$ magnitude $V_{\sim n}$ begins to increase, and the vertical portion of the I–V characteristic of the normal mode starts to move in the opposite direction. This situation is used in the known method of 'hindered discharge' to determine the parameters of the cathode fall of the DCGD [159].

For completeness, we give the form of the current–voltage characteristics obtained for other types of RFCD: 1) the current–voltage characteristic with external electrodes for various insulating dielectrics (Figs. 1.22, 1.23); 2) CVC with a single internal electrode (Fig. 1.24).

As can be seen in Fig. 1.22, in the case of insulation of electrodes with dielectrics with a small value of the dielectric constant ε, the form of the current–voltage characteristic is practically independent of the gas filling of the discharge tube (the gas pressure varied by three orders of magnitude). This indicates that the properties of the resulting discharge are entirely determined by the impedance of the insulating dielectric, and not by the impedance of the discharge itself. In this case, the CVC does not completely correspond to that for the

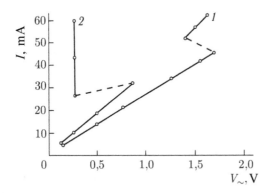

Fig. 1.24. CVC in the active electrode circuit. Air, $p = 15$ Torr. $f = 1.2$ MHz, $d = 5$ mm. *1* – single-electrode discharge, *2* – two-electrode discharge.

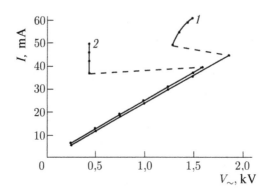

Fig. 1.25. CVC in the active electrode circuit. Air, p = 20 Torr, f = 1.2 MHz, d = 32 mm. *1* – single-electrode discharge, *2* – two-electrode discharge.

case of RFCD with internal electrodes. A completely different form is shown in Fig. 1.23 CVC using a dielectric with an anomalously large value of ε, which has a small impedance. In this case, the current–voltage characteristic of a discharge with external electrodes is very close in appearance to the I–V characteristic of a discharge with internal electrodes, determined by the gas filling of the tube.

As was experimentally established, the RFCD burns in the discharge tube and in the presence of only one internal electrode. For comparison, Fig. 1.24, 1.25 are the CVCs obtained in one tube for two-electrode and single-electrode discharges. In the latter case, one of the electrodes was not connected to a power source, nor was it 'grounded', but simply 'suspended in the air'. The experimental conditions of Fig. 1.24, 1.25 differ only in the distances between the internal electrodes.

Comparison of the I–V characteristics of the two-electrode and single-electrode discharges shows that their lower branches (gas-discharge α-branches or purely bias current) behave qualitatively identically for both discharges, but the behaviour of the upper branches is very different. If in the two-electrode RFCD the upper branch is a typical γ-branch of the discharge in the 'normal' mode, then in the one-electrode discharge the upper branch behaves similarly to the rectilinear lower branch, differing only in the large value of the angle of inclination.

The behaviour of the I–V characteristic of a single-electrode discharge seems quite natural if one takes into account the fact that there is no electric circuit for this RFCD to pass the conduction current, and therefore only the bias current exists

$$I_\sim = C\varpi V_\sim,$$

where C is the effective capacity of the discharge gap. The increase in the angle of inclination of the linear portion of the upper branch after gas breakdown is due to the appearance of a space charge in the plasma tube and layers near the existing surfaces, which increase the effective capacity of the discharge gap C and, respectively, the discharge current $I_\sim$. Other conclusions from the comparison of the CVCs presented in Figs. 1.24 and 1.25 will be given below.

Summarizing the results of the experimental study of the CVC of the RFCD one can defined four types of CVC characteristics differing in the physical sense: 1) current–voltage characteristic of a true two-electrode RFCD, 2) separate current–voltage characteristics in circuits of active and grounded electrodes of RFCD, 3) CVC with external electrodes, 4) current–voltage characteristic of single-electrode RFCD.

Using the measurements described above (Fig. 1.19 *b*), we can construct a curve for the dependence $V_{\sim n}(d)$ and obtain the spatial structure of the RFCD, which will allow us to determine the values of d_{ns}, $V_{\sim ns}$, and the RF field strength $E_\sim$ in the positive-column plasma. The curve of the dependence $V_{\sim n}(d)$ will be presented below.

On the basis of the procedure for obtaining a family of CVC of the RFCD and analysis of the latter with the use of a justified equivalent electric circuit of this category it was possible to propose a new method for diagnosing RFCD and its plasma – 'CVC – method of diagnosing RFCD'.

1.6. Modern ideas about the classification of RFCD by types

Based on the literature results of studies of the physical properties of RFCD, it seems that at present it is quite timely to raise the question of a new classification of RFCDs by types.

It has been known for more than 50 years [6, 23] that RFCD can exist in two modes: low-current (α-discharge) and high-current (γ-discharge). In the case of medium gas pressures, the two discharges mentioned above strongly differ both in appearance and in the nature of the physical processes in the NESCL. At low pressures, there are also α- and γ-discharges, however, the difference between them is no longer manifestly clear.

In the α-mode, electrons acquire energy in the RF field in a quasi-neutral plasma, while the emission of electrons from the surface of

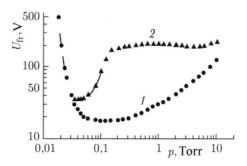

Fig. 1.26. The curve of the extinction of the RF discharge *1* and the curve of the α–γ transition *2* at *d* = 2.2 mm [161].

the electrodes in maintaining the discharge does not play a special role. In this case, the conductivity of the NESCL is small, and the current between the plasma and the electrodes is practically closed by the bias current.

In the case of the γ-mode, electron avalanches develop in the NESCL, and gas ionization by the electron impact gas occurs mainly near the NESCL–plasma boundary, while the emission of electrons from the surface of the electrodes significantly affects the process of electron multiplication and maintenance of the discharge. The NESCL in the γ-mode have a significant conductivity and are similar in their properties to the cathode layer of the DCGD.

As was experimentally established [160], the region of stable existence of the α-mode is limited by the middle pressures: for a fixed distance between the electrodes *d*, there exists a pressure p_{cr} such that for $p \geq p_{cr}$ the discharge can only burn in γ-mode. This is very clearly represented when comparing the experimental discharge extinction curves and the $α \rightarrow γ$ transition obtained in [161] (Fig. 1.26). It is also clear from this that, at a sufficiently low pressure (*p* < 0.03 Torr), the region of existence of the α-mode is limited by low pressures, where the RFCD can only burn in γ-mode. In this case, the γ-mode of the discharge is visually very similar to the α-mode, which can easily be misleading.

One more difference in the properties of α- and γ-discharges, due to their different physical mechanisms, should be noted. In the γ-discharge at medium pressure the plasma in the centre of the discharge gap has a high electronic temperature close to the temperature T_e in the α-discharge. At low pressures, T_e in the plasma of the γ-discharge is very low, more than an order of magnitude lower, compared to a similar temperature in the α-discharge.

Experiments [160] and calculations [161] show that RFCD at medium pressures passes abuptly from α- to γ-mode, while the plasma density and the amplitude of RF current increase several times. This is also seen in Figs. 1.18 and 1.19. The transition to the same is accompanied by a complete rearrangement of the spatial structure of the discharge.

At intermediate ($p \sim 1$ Torr) and low pressures ($p < 0.1$ Torr) in the RFCD, the $\alpha \rightarrow \gamma$-transition occurs smoothly, without jumps.

In the monograph in [6], the $\alpha \rightarrow \gamma$-transition is interpreted as the result of breakdown of the layers of space charge of the α-discharge. In this case, the ionic NESCL is conditionally considered as an electron-free (not 'punctured') gas gap with the thickness average for a period of a field d_α. As the density of the discharge current increases, the density of ions in the layer increases, which leads to an increase in the average voltage drop across layer $\bar{V}$ to a value corresponding to the breakdown threshold of a gas gap of the same length d_α. The values of $\bar{V}$ and pd_α approximately correspond to a point on the Paschen curve for such an interval. The breakdown condition is formulated as the condition of Townsend's self-maintenance of the current of charges in the layer averaged over the RF field period due to ion–electron emission from the electrode with subsequent charge multiplication.

After the NESCL breakdown, an ionization wave passes to the electrode, which leaves a plasma behind it and shortens the original layer of thickness d_α to a thickness d_γ corresponding to optimal conditions for charge multiplication. This process is similar to the process of formation after a breakdown of a DCGD with a 'normal', reduced layer thickness d_n.

The described process occurs at medium pressures, when the pd_α value for the α-discharge corresponds to the right branch of the Paschen curve. At low pressures, when pd_α corresponds to the left branch of the Paschen curve, the formation of a 'normal cathode layer' would require not a contraction, but an extension of the NESCL. A detailed analysis of the physical processes under these conditions, carried out in [6], showed that this is impossible. Therefore, in a low-pressure discharge, the $\alpha \rightarrow \gamma$-transition occurs without jumps in the thickness of the near-electrode layer and other parameters. At the same time on a continuous CVC, the RFCD only appears as a kink.

In [161], the question of the formulation of the criterion for the $\alpha \rightarrow \gamma$-transition to RFCD at low and intermediate gas pressures is considered. It is noted that, on the one hand, the voltage of the $\alpha \rightarrow \gamma$ transition is represented by the value $V_{\sim}$ at which a jump of the derivative $dI_{\sim}/dV_{\sim}$ [128, 163] is observed. On the other hand, it was assumed in Ref. [23] that at low pressures the $\alpha \rightarrow \gamma$-transition curve coincides with such a section of the discharge ignition curve that, after the gas breakdown, the RFCD burns immediately in the γ-mode. We recall also the criterion of the $\alpha \rightarrow \gamma$-transition cited earlier [6] – the appearance of breakdown of near-electrode layers of the α-discharge in accordance with the Townsend condition should be taken into account.

The above criteria often do not agree with each other and contradict the experimental data. For example, from the data given in [161, 164], it can be concluded that, with a decrease in the Ar pressure, the voltage required for the transition from the α- to γ-mode is lower than the breakdown of the RF voltage. The verification of the fulfillment of the Townsend breakdown criterion in the $\alpha \rightarrow \gamma$ transition, performed in Ref. [161], showed that this criterion is not completely satisfied. Indeed: at the pressure of Ar of $p = 0.1$ Torr in the experiment, the transition occurred at a voltage $V_{\sim} = 85$ V, and to fulfill the Townsend criterion it would be necessary to apply an RF voltage $V_{\sim} = 850$ V. Thus, in the wide range of RF voltages in the γ-mode, the near-electrode layer is not penetrated. In addition, visual observations show that at low and intermediate pressures, the structure of RFCD becomes very similar to the structure of DCGD, when the voltage applied to the electrodes of the RF is still clearly insufficient for breakdown of the near-electrode layers.

In view of the foregoing, the following change in the RFCD parameters was adopted in [161] as a criterion for the transition from α- to γ-mode. At Ar pressures $p > 0.5$ Torr, the transition $\alpha \rightarrow \gamma$ occurs without jumps, and it was believed that the $\alpha \rightarrow \gamma$ transition occurs at an RF voltage at which the conduction current in the electrode circuit $I_{\sim}\cos\varphi$ and the charge density of the plasma n_i in the centre discharge reach a maximum. At pressures $p = 0.05$– 0.5 Torr the $\alpha \rightarrow \gamma$-transition is accompanied by a sharp rearrangement of the discharge glow structure and a sudden decrease in the plasma density n_i and the electron temperature T_e in the central region of the discharge with increasing phase angle φ between the current and voltage.

We note one more observed effect accompanying the $\alpha \rightarrow \gamma$-transition. In [161], the dependence of the conduction current in the electrode circuit $I_{\sim}\cos\varphi$ on the RF voltage $V_{\sim}$, shown in Fig.

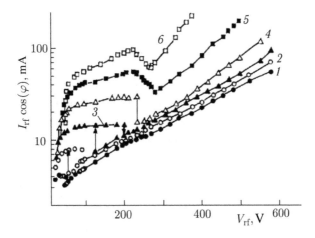

Fig. 1.27. The dependence of the RF conduction current in the electrode circuit on the applied RF voltage. $d = 22$ mm, argon pressure p, Torr: *1* – 0.06, *2* – 0.1, *3* – 0.2, *4* – 0.5, *5* – 1, *6* – 2.

1.27, was experimentally studied in a rather wide range of pressures ($0.06 \leq p \leq 2$ Torr).

The curves $I_{\sim} \cos\varphi$ $(V_{\sim})$ show that at argon pressures $p \leq 0.5$ Torr, the jumplike $\alpha \rightarrow \gamma$ transition has a hysteresis nature. The latter is explained as follows. The electrons emitted from the electrode, being accelerated in the NESCL, form a high-energy near-electrode electron beam. Since in this pressure range the ionization length of the electron path λ_i exceeds the layer thickness d_s, a significant part of the ionizing collisions will occur not only near the layer boundary, but also in the near-electrode plasma. In terms of its characteristics, this plasma, formed by processes from two electrodes, is analogous to two regions of negative luminescence (NL) and two regions of Faraday dark space (FDS) of a DC glow discharge, overlapping in the centre of the RFCD under consideration. As is well known [4], in these parts of the discharge the electric field is small. Consequently, in the $\alpha \rightarrow \gamma$ transition the RF voltage drop on the NESCL increases, and the inverse $\gamma \rightarrow \alpha$ transition occurs at lower RF voltages, creating a hysteresis.

At higher pressures ($p > 0.5$ Torr) at the beginning of the the $\alpha \rightarrow \gamma$ transition, two FDSs do not overlap, and a plasma section of the PC is observed in the centre of the discharge. An increase in RF voltage leads to an increase in the length of the FDS, the occupied PC region narrows, and at higher RF voltages two FDSs overlap at the centre of the discharge. Thus, now the RFCD turns into the γ-mode continuously, and for the inverse $\gamma \rightarrow \alpha$ transition, the hysteresis disappears. The foregoing shows that the physical properties of the α- and γ-type discharges have been fairly well studied.

As is well known [6], electrons in the RF discharge are heated as a result of collisions with atoms, when the energy of the ordered oscillatory motion in an alternating field passes into chaotic energy after each collision. In this case, the Joule heat acquired by the electrons is proportional to the frequency v_{en} of electron-atom collisions. It follows that at very low pressures a very small power must be released in the discharge. However, reliable experiments under these conditions have shown the release of a power that is more than one order of magnitude expected for collisional heating. The large discrepancy found is explained by the existence of an alternative mechanism for heating electrons in the RF discharge, which also acts in the absence of collisions between the particles.

In the early 1970s, a hypothesis was advanced on the possibility of additional heating of the plasma electrons of low-pressure RFCDs by an oscillating boundary of the NESCL [50, 51, 56]. In the literature this mechanism is called stochastic heating. It was stressed [50, 51] that this is a new discharge maintenance regime that is different from the traditional α- and γ-combustion modes of the discharge. In the terminology of [168], such a regime can be called 'electron-layer-collisional', which can be designated as the s-mode from the English word *sheath.*

The next regime for the existence of low-frequency RFCDs – the 'RF–beam–plasma discharge' (RF BPD) was experimentally observed in [145, 169], when a strongly asymmetric discharge with a small active electrode diameter (~5 mm) and a large area of the grounded electrode, as well as with high RF voltages ($V_{-} \geq 1$–2 kV) was studied. The original feature of this discharge was a substantial increase in the density of the near-electrode electron beams, which provided the excitation of the beam-plasma instability (BPI) in a gas-discharge plasma [155, 169, 145]. The new type of discharge obtained from the monotonous increase in RF voltage from the γ-type of the RFCD – RF BPD – is characteristic by the presence of a 'plasmoid' in its spatial structure with an anomalously high electron temperature $T_e \geq 5 \cdot 10^6$ K and other attributes of the classical BPD, traditionally obtained with the help of electron guns, generating high-energy electron beams [170, 171]. In more detail, the properties of

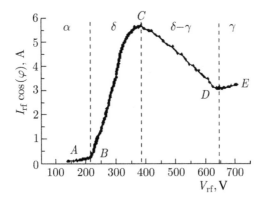

Fig. 1.28. Dependence of the RF current on the applied RF voltage for SF_6 pressure $p = 0.4$ Torr and the electrode spacing $d = 20.4$ mm [172].

RF BPD, which can be designated as '*i*-discharge' (from the English word *instability*), will be described below.

It is quite appropriate to be present as a new independent regime the RFCD described in [172] as the 'dissociative' regime of low-density RFCD in a number of molecular gases. In this paper, the results of an experimental study of RFCD in SF_6, NF_3, and SiH_4 are presented.

It is shown that RCCD in these gases can exist not only in α- and γ-modes, but also in the 'dissociative' δ-mode (from the Greek word '*διασπαση*' – dissociation). This δ-mode is characterized by a high degree of dissociation of gas molecules, high plasma density, electron temperature and active discharge current. This δ-mode is intermediate between α- and γ-modes. The reason for the appearance of the δ-mode is a sharp increase in the dissociation rate of molecules by electron impact, beginning with a certain threshold value of the RF voltage $V_{\sim}$. For the radicals formed (SF_x $x = 1$–5, NF_x $x = 1$–2, SiH_x $x = 1$–3), the threshold ionization energy is less than the ionization potential of the initial SF_6, NF_3, and SiH_4 molecules. The current–voltage characteristic of the RFCD in the presence of a 'dissociative' mode is shown in Fig. 1.28.

It is shown that the double layer existing in the 'anodic' phase of the NESCL plays an important role in maintaining both the α-mode and the δ-mode, but it is not the cause of the $\alpha \rightarrow \delta$ transition of RFCD. Additional data on the physical properties of RFCD in the 'dissociative' mode will be given below.

Finally, in principle, one should mention such a poorly known form of the RFD as this discharge at high frequencies of the RF field of 500–900 MHz, when even in the plane geometry of the electrodes the discharge is supported not only by the potential, but also by the vortex components of the RF electric field. If, theoretically, papers on this subject have started to appear [173], in the experimental respect such an RFD is practically not studied. It seems natural to consider the discharge, simultaneously supported by RF potential and vortex electric fields, as an independent, 'hybrid' mode of the RFD – 'g-mode'.

On the basis of what was said earlier, it is possible to propose a method to refine the traditional division of RFCDs only into α- and γ-discharges and propose a new classification for RFCD:

1) traditional α-discharge;

2) traditional γ-discharge;

3) the s-discharge with stochastic heating of plasma electrons by oscillation of the boundary of the near-electrode layer;

4) RF beam-plasma discharge or i-discharge;

5) discharge in dissociative mode or δ-discharge;

6) a hybrid discharge at increased radiofrequencies or a g-discharge.

1.7. Anharmonicity of the discharge current of the RFCD

The elements of the RFCD mechanism are reflected, in particular, in the time course of the discharge current during one period of the RF field. As is known, with a harmonic voltage of a high-frequency power source, the discharge current can, depending on the experimental conditions, both follow in form with the voltage, and be in different degrees anharmonic. The experimental detection of higher harmonics in current oscillograms was noted quite a long time in a review paper [174].

Earlier, attempts were made to find the causes of current anharmonicity in a number of theoretical and experimental studies [78, 175, 176] and the monograph [6]. In this case, a very simplified electrotechnical model of the RFCD was used, in which they operated with some average capacitances and thicknesses of the electrode layers, not taking into account the nature of their real behaviour in time.

As is known, an electric circuit with constant capacitances is a linear system, and with a sinusoidal voltage on the electrodes, the discharge current must also vary according to the harmonic law.

Actually, the approximately sinusoidal form of current is to some extent peculiar only to symmetrical discharges with identical electrodes of equal area $A_1 = A_2$ and constant ion density in the near-electrode layers with thicknesses d_{s1} and d_{s2}. In this case, the equivalent capacity of two layers C_{eq}

$$C_{eq} = \left(C_1^{-1} + C_2^{-1} \right)^{-1} = \left(\frac{4\pi d_{s1}}{A_1} + \frac{4\pi d_{s2}}{A_2} \right)^{-1}$$

is constant because of the constancy of the sum of the time-varying thicknesses:

$$d_{s1} + d_{s2} = \text{const},$$

and the condition for the constancy of the ion density in the NESCL is satisfied only in discharges at gas pressures not lower than the average.

In the asymmetric RFCD ($A_1 \neq A_2$), the following holds: the value of the capacitance C_{eq} varies over a wide period of the RF field, the

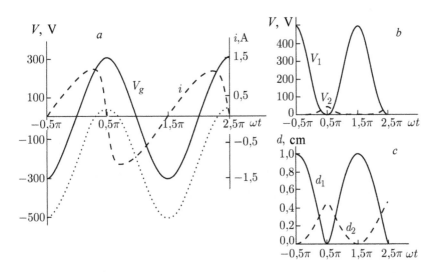

Fig. 1.29. The evolution of asymmetric discharge parameters, calculated on the basis of the simplest model: the voltage of the RF generator V_g, the potential of the small electrode V_1, the discharge current i [6].

laws of motion of the boundaries of NESCL, the voltage drops on them, and the discharge current are anharmonic even for the purely sinusoidal nature of the applied voltage. In [6] this is treated as follows.

Due to the asymmetry of the discharge, the plasma boundary changes its direction of motion at the moments of contact with the small electrode much faster than at the moments of contact with the large electrode. This is caused by the need to provide the displacement current density at an electrode of a smaller area higher than for a large electrode. This cause determines the nature of the current change during the field period. The current varies relatively rapidly from peak to peak at the moments of proximity of the plasma to the small electrode, and then comparatively slowly changes to the next peak. Therefore, instead of a symmetrical sinusoid, the calculated current oscillogram acquires a 'sawtooth' character, as seen in Fig. 1.29.

Note that in a low-pressure symmetric RFCD ($p < 10^{-1}$ Torr), some anharmonicity of the current is also observed, but the cause and nature of the anharmonicity are different: anharmonicity is caused by the inconsistency of the ion density in the NESCL [6].

Thus, according to the literature, the cause of the anharmonicity of the current RFCD is two factors: 1) the asymmetry of the electrodes and 2) the inhomogeneity of the ion density in the NESCL. In this case, often both mentioned factors act simultaneously, and it is very

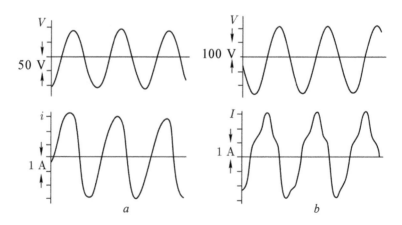

Fig. 1.30. Experimental oscillograms of the sinusoidal voltage and current in a discharge between coaxial cylinders with radii of 5 and 12.5 cm (Ar, $p = 3 \cdot 10^{-3}$ Torr, $f = 13.56$ MHz): *a*) with a longitudinal magnetic field; *b*) without a magnetic field [177].

difficult to determine experimentally the relative contribution of each of them.

So far, only one work [177] has been devoted to the study of RF of the magnetron discharge; in the authors' opinion, the obtained experimental data allow one to see the change in the nature of the anharmonicity of the current when one of its causes is eliminated. However, it appears that the elimination of the inhomogeneity of the ion density in the NESCL by applying a longitudinal magnetic field to the discharge is not a completely correct method in the sense of a simplified interpretation of the consequences of its application. In this case, the nature of the motion of charged particles in the radial direction with different degrees of magnetization of electrons and ions must change substantially.

There is little experimental data on the nature of the anharmonicity of the current in various physical conditions. In this case, the oscillograms of the total discharge current I_t, representing at each instant of time the sum of its reactive I_r and the active I_a constituents:

$$I_t(t)=I_a(t)+I_r(t).$$

The oscillogram of the current $I_t(t)$ is difficult to interpret without taking into account the time course of its components $I_a(t)$ and $I_r(t)$. Of greatest interest is the experimental study of the anharmonicity of the active current $I_a(t)$, which is associated with the motion of charges in the NESCL in certain phases of the RF field period. With the appropriate development, such studies could serve as a basis for

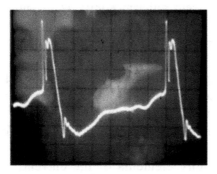

Fig. 1.31. Oscillogram of the discharge current of a highly asymmetrical RFCD. Ne, $p = 0.5$ Torr, $f = 1$ MHz, $V_\sim = 500$ V, active electrode diameter – 5 mm, grounded electrode – 60 mm.

the creation of new, contactless methods for diagnosing physical processes in the NESCL of the RFCD.

In the cases of strong anharmonicity of the $I_t(t)$ behaviour, the observed oscillations of the current type, which are quite complicated in form, are shown without explanation [117] (Fig. 1.30) and presented below are even more complicated.

It should be noted that the currently available literature studies of the nature of the anharmonicity of the current of the RFCD are based only on a very approximate electrotechnical discharge scheme as a whole, which can give only some elements of the anharmonicity mechanism commensurate with time intervals of the order of the period of the RF field T.

The main physical processes described in [50, 67, 169], which can substantially influence the dynamics of the discharge current of RFCD, practically do not fall in the field of view of the researchers. Some idea of the complexity of the processes occurring in the discharge can be given by an oscillogram of the current of a highly asymmetric low-pressure discharge, shown in Fig. 1.31.

The above oscillogram shows the presence within the RF field period of short-term non-monotonic changes in the current with a duration τ that is substantially shorter than the RF field period, $\tau \ll T$. It will be shown below that the observed fine structure of the temporal behaviour of the current is associated with short-term 'flashes' of excitation by pulsed electron beams of beam-plasma instability (BPI) [169].

The state of affairs in literature today with the explanation of the anharmonicity of the current of the RFCD reflects the statement made in [6]: 'Using electrical engineering terminology, we can say that the cause of current anharmonicity is the non-linearity of the layer resistance.' Considering what was said earlier, one can note that this explanation is partly correct, but not quite concrete and far from complete.

To further advance along the path of a satisfactory explanation of the physical nature of the phenomenon under consideration, it is necessary to experimentally study the anharmonicity of the discharge current of RFCD in various physical conditions, in particular the active component of the current, and a detailed analysis of the observed anharmonicity elements of the current, using previously obtained data on the physical mechanism of RFCD for elucidating the possibility of existence mechanisms of the appearance of the anharmonicity of current, alternative ones that are given in the literature.

1.8. Spatial structure of RFCD

The spatial structure (SS) of the gas discharge, which gives a definite idea of the physical conditions at each point of the volume between the discharge electrodes, is a very informative characteristic of the discharge properties. In a certain sense, the SS is a physical discharge mechanism deployed in space.

Usually, the SS of the integral glow of the discharge is studied. However, it is obvious that a rather complete picture of the physical conditions in the discharge gap, in our case of the RFCD under investigation, gives only the spatial distribution of a large number of gas discharge characteristics and parameters, such as *the intensity of the integral luminescence I_c, the intensity of the individual spectral*

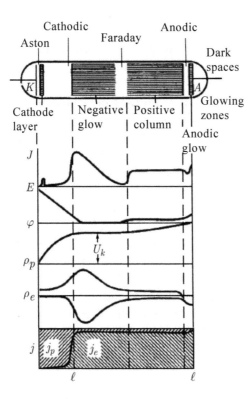

Fig. 1.32. The structure of the DCGD and the distribution of the luminescence intensity I, the field strength E, the potential φ, the charge densities, and the currents of the positive ions and electrons ρ_p and ρ_e, j_p and j_e [151].

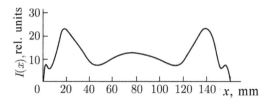

Fig. 1.33. The spatial distribution of the integrated luminescence intensity $I(x)$ in the RFCD of the γ-type. He, p = 0.25 Torr, f = 4 MHz, $V_\sim$ = 400 V.

lines I_λ, the total optical emission spectrum $S(\lambda)$, the amplitude of the oscillations of the integrated radiation and individual spectral lines at the fundamental frequency of the RF field f and its second harmonic 2f – I_f, I_{2f}, $I_{\lambda f}$, $I_{\lambda 2f}$, the degree of polarization of the radiation with spectral lines P_λ, RF electric field strength $E_\sim$, quasi-stationary electric potential φ, concentration of charged particles n_e, electron temperature T_e or electron energy distribution function $f_e(\varepsilon)$, ionization frequency v_i and others.

For illustration, we give the spatial distributions of only some characteristics of RFCD.

At one time, to study the physical properties of DCGD, much information was provided by a detailed experimental study of its SS [151, 182], shown in Fig. 1.32.

The presence of all spatial regions presented in Fig. 1.32 depends on the pressure p and the interelectrode distance d. Bridging the electrodes, you can remove all the areas, except for the area of cathode drop required to maintain the DCGD. The terminology that arose in relation to the areas of the SS of the DCGD was later applied to describe RFCD.

As has been established experimentally by many researchers, the structure of the near-electrode luminescence of the RFCD of the γ-type coincides with that of the near-cathode region of DCGD. A typical SS of a symmetric RFCD of the γ-type is shown in Fig. 1.33. One can see a complete analogy with two cathodic inclusions of the DCGD turned on opposite to each other, where the plasma of the central discharge region serves as an effective anode.

The distributions of the integral luminescence $I(x)$ in the near-electrode region of the RFCD and the cathode region of the DCGD when the discharge is ignited in the same discharge tube are shown in Fig. 1.34.

Figure 1.35 shows the change in the character of the spatial distribution of the intensity of the integral luminescence in the near-electrode region of the RFCD with an increase in the RF voltage $V_\sim$.

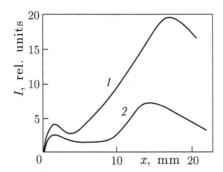

Fig. 1.34. The spatial distribution of the intensity of the integral luminescence in the near-electrode region of RFCD and the cathode region of DCGD in the same discharge tube. *1* – RFCD Ne, *p* = 0.25 Torr, $V_\sim$ = 500 V, *f* = 4 MHz, electrodes –Ti. *2* – DCGD Ne, *p* = 0.25 Torr, *V* = 700 V, *f* = 4 MHz, electrodes – Ti.

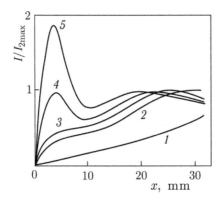

Fig. 1.35. Dependence of the normalized intensity of the integral brightness in the near-electrode region of the RFCD on the 2nd maximum on the amplitude of the applied RF voltage $V_\sim$. Ne, *p* = 0.25 Torr, *f* = 2 Mc. $V_\sim$: *1* – 100 V, *2* – 300 V, *3* - 600 V, *4* – 800 V, *5* – 1000 V.

It is seen that at low $V_\sim$, the emission intensity monotonically decreases from the centre of the discharge to the electrode, which corresponds to the predominant contribution of the volume mechanism of ionization and due to the RF field ('α-discharge'). As the RF voltage rises, a luminescence distribution *I* (*x*) is formed, identical to that in the DCGD, and the transition of RFCD to the γ-mode takes place.

Let us briefly consider the spatial nature of RF and quasi-stationary electric fields in the discharge gap of the RFCD.

A typical form of the distribution of the strength $E_\sim$ of the RF field obtained with the aid of the triple-cylindrical probe technique [183] is shown in Fig. 1.36. In this case, a symmetrical RFCD with

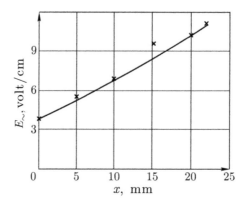

Fig. 1.36. The distribution of the RF field strength $E_\sim$ along the discharge gap in the RF discharge He, $p = 0.48$ Torr, $f = 6$ MHz, $V_\sim = 600$ V, $d = 6$ cm.

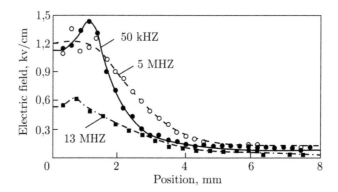

Fig. 1.37. The dependence of the RF field strength $E_\sim$ on the distance to the active the RFCD electrode in the negative half-cycle of the high-frequency field [185].

an interelectrode distance $d = 6$ cm was investigated. Because of the relatively small spatial resolution of the triple probe method, measurements were made in the central region of the discharge, where there was a relatively slow change in $E_\sim$, and ceased at a distance $\Delta x = 0.8$ cm from the electrode in regions of an increased gradient of the field strength. Thus, the technique used made it possible to find only spatially averaged values of the probe $E_\sim$.

The spatial distribution of the RF field strength $E_\sim (x)$ in RFCD was investigated with high spatial and temporal resolution by the modern method of laser-induced fluorescence (LIF) in [184, 185]. The results of these measurements of $E_\sim (x)$ for different frequencies of the RF field are shown in Fig. 1.37 and will be considered below.

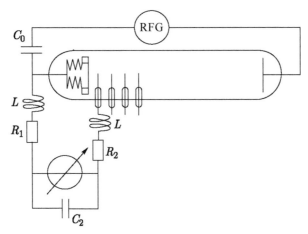

Fig. 1.38. Electrical circuit with a single cylindrical probe for measurements of the spatial distribution of the quasi-electrostatic potential $\varphi(x)$ and the jump of the potential U_s in the NESCL of the RFCD.

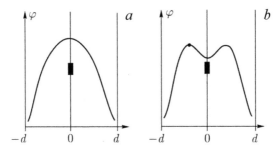

Fig. 1.39. A characteristic form of the spatial distributions of the quasi-stationary potential $\varphi(x)$ in the RFCD: a) $p < 1$ Torr; b) $p > 1$ Torr.

The spatial distribution of the quasi-electrostatic potential $\varphi(x)$ was carried out using a single Langmuir probe. In this case, one mobile probe or several probes (4 or 5) were used, which were soldered into the wall of the discharge tube with an interval of several millimeters. Smooth movement of the discharge electrode relative to the probes was made with the aid of a micrometer screw (Fig. 1.38).

Qualitatively, the distributions $\varphi(x)$ of two types were observed: a) for pressures $p <1$ Torr; b) for pressures $p > 1$ Torr (Fig. 1.39). The possibility of a non-monotonic behaviour of the potential must be taken into account when measuring the jump in the potential in the NESCL.

A characteristic form of the spatial distribution of the quasi-stationary electric potential $\varphi(x)$, observed in a symmetrical RFCD

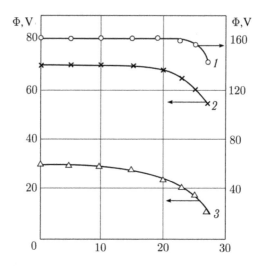

Fig. 1.40. Distribution of the quasi-stationary potential of the electric field in the discharge gap of RFCD He, p = 0.5 Torr, f = 5 MHz. $V_\sim$: *1* – 300 V, *2* – 120 V, *3* – 40 V, d = 6 cm.

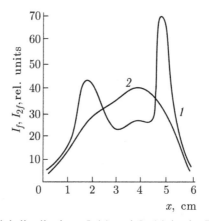

Fig. 1.41. The spatial distributions $I_f(x)$ and $I_{2f}(x)$ in the RFCD (active electrode on the right, grounded to the left) He, p = 1 Torr, f = 1.2 MHz, $V_\sim$ = 150 V.

with a short discharge gap, is shown in Fig. 1.40. As can be seen, the central region of the discharge gap with increased RF voltage is practically an equipotential space. This can be explained by the fact that in the short discharge gap two regions overlap in the centre – analogues of the regions of the 'negative glow' (NG) of the DCGD from the opposite electrodes. The regions of the NG are, as is well known, equipotential [151].

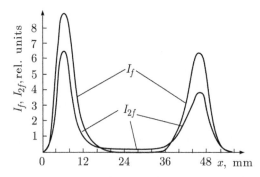

Fig. 1.42. The spatial distribution of the amplitudes of the luminescence oscillations I_f and I_{2f} in the RFCD of medium pressure He, $p = 10.4$ Torr, $f = 1.2$ MHz, $V_{\sim} = 850$ V.

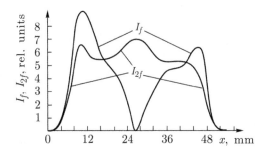

Fig. 1.43. The spatial distribution of the amplitudes of the luminescence oscillations I_f and I_{2f} at RFCD at high RF voltage. He, $p = 5.8$ Torr, $f = 1.2$ MHz, $V_{\sim} = 1700$ V.

It will be shown below that, in low-pressure RFCD with high RF voltages ($V_{\sim} \geq 200$ V), there are two factors affecting the plasma: the RF field and the near-electrode electron beams (NEEB). At each point of the discharge volume, the RF field heats the electrons twice during the period of the field, which causes oscillation of the luminescence of the plasma with a doubled frequency $2f$.

At the same time, as shown below, the NEEB in RFCD are pulsed and exist only in the negative half-cycle of the RF field. Thus, the NEEB affect the plasma with the frequency of the RF field f.

By recording in space the amplitudes I_f and I_{2f} of the components of the luminescence oscillations at the frequencies f and $2f$, it is possible to establish qualitatively in the discharge gap the regions and the relative intensity of the manifestations of the plasma-exciting factors.

The characteristic distributions of $I_f(x)$ and $I_{2f}(x)$ in the discharge gap for different experimental conditions (gas pressure p and amplitude of RF voltage $V_{\sim}$) are shown in Figs. 1.41–1.43.

The obtained distributions $I_f(x)$ and $I_{2f}(x)$ show the presence of spatial regions in the discharge space of the RFCD, where the effects of electron beams and the RF field are substantially manifested, respectively. In more detail these physical processes are considered in the future, and now only some features of the spatial picture of the distributions $I_f(x)$ and $I_{2f}(x)$ are noted here.

With the increase in pressure, even at the beginning of the range of middle pressures, in spite of the significantly increased RF voltage, both beam and RF field effects are concentrated in the near-electrode regions (Fig. 1.42).

The most complex picture of the spatial distributions $I_f(x)$ and $I_{2f}(x)$ is observed at high RF voltage, when a significant increase in the role of electron beams in the discharge maintenance mechanism should be expected (Fig. 1.43). First of all, we should pay attention to the following peculiarities in the behaviour of the distributions obtained: 1) a sharp decrease to zero at the centre of the discharge of the amplitude of the first harmonic I_f, 2) the appearance in the central region of the third maximum in the distribution of the 2nd harmonic I_{2f}.

Postponing further detailed analysis of the results obtained, a significant methodological observation should be made immediately with respect to the measurements. The point is that at high RF voltages, high-energy counterpropagating electron beams from opposing electrodes can cover the entire discharge gap, especially its central part. Since in each period of the RF field these pulsed beams exist in antiphase, in the central region of the discharge,

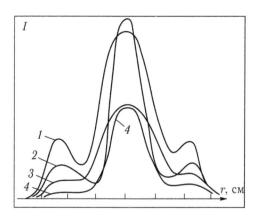

Fig. 1.44. Dependence of the radial distributions $I_f(r)$ and $I_{2f}(r)$ on the diameter of the RFCD electrodes. Xe. $p = 1$ Torr, $f = 1.2$ MHz, $V_\sim = 1250$ V. $1 - I_f(r)$, $D_{el} = 20$ mm; $2 - I_f(r)$, $D = 5$ mm; $3 - I_{2f}(r)$, $D = 20$ mm; $4 - I_{2f}(r)$, $D = 5$ mm.

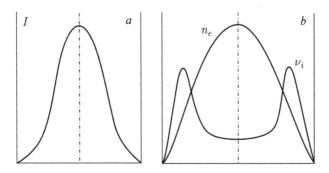

Fig. 1.45. Spatial distribution of parameters in the α-type RFCD: *a*) the distribution of luminescence intensity $I(x)$; *b*) the distribution of $n_e(x)$, $v_i(x)$.

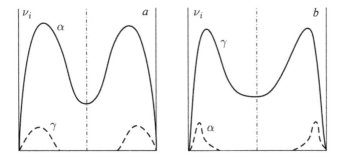

Fig. 1.46. The spatial distribution of ionization frequency $v_i(x)$ in the RFCD for different RF voltages. He, $p = 1$ Torr. *a*) $V_\sim = 120$ V; *b*) $V_\sim = 400$ V. α – thermal electrons, γ – beam electrons.

overlapping it in turn, they can make a significant contribution to the amplitude of the signal at the second harmonic. The latter may appear as a false enhancement of the RF field effects in this region, which should be interpreted correctly.

The above distributions $I_f(x)$ and $I_{2f}(x)$ show the behaviour of these characteristics in the discharge gap of the RFCD in the longitudinal direction. We also investigated the distributions of the amplitudes I_f and I_{2f} along the radius of the discharge tube. The typical radial distributions $I_f(r)$ and $I_{2f}(r)$ obtained are shown in Fig. 1.44.

The radial distributions $I_f(r)$, $I_{2f}(r)$ will be discussed below.

The simplest SS has a symmetric RFCD of the α-type of planar geometry in the ambipolar diffusion regime (Fig. 1.45). The distribution of the luminescence intensity in it is bell-shaped, smoothly falling to the electrodes (Fig. 1.45 *a*), the charge density has a cosine distribution $n_e(x) = n_{e0}\cos(\pi x/d)$, the ionization frequency distribution $v_i(x)$ has maxima in the near-electrode regions where the RF field is concentrated (Fig. 1.45 *b*).

The spatial character of the ionization processes (ionization frequency v_i) obtained by the numerical method in RFCD with small and high RF voltage $V_\sim$ is qualitatively represented in Fig. 1.46.

Here, the distribution curves $v_i(x)$ in the discharge gap are denoted by the indices α and γ, which means ionization by low-energy thermal electrons (α) and high-energy beam electrons (γ). As can be seen from the curves $v_i(x)$ in Fig. 1.46 a, Fig. 1.46 b, for small $V_\sim$ ionization is mainly produced by α-electrons, and for large $V_\sim$ by γ-electrons.

According to what was said earlier, it is possible to determine the type of RFCD (α- or γ-discharge) purely visually by the nature of the luminescence, and in the case of the γ-discharge its modes (normal or anomalous).

Taking into account the available data on the SS of the RFCD, it is possible to purposefully organize a discharge for the practical application of the plasma of its various spatial regions, for example, the plasma of the 'positive column' or 'negative glow'.

1.9. Dependence of the physical properties of RFCD on the configuration of the discharge gap

Numerous practical applications of RF discharges are associated with the use of different discharge gap configurations. The technical methods for realizing the RF discharge are much more diverse than the discharge of direct current. As a result of the conducted experimental studies it is established that the discharge configuration can affect not only its individual quantitative characteristics, but also the mechanism of the RF discharge [145]. It should be noted that the literature does not pay enough attention to this issue. Meanwhile, the importance of choosing the configuration of the electrodes that excite the RF discharge is already evident from the fact that at one of them a capacitive RF discharge appears, and at the other, an inductive one. Even the physical nature of the fields creating these discharges is qualitatively different – this is the potential and vortex fields, respectively.

For a more precise exposition of the main dependences of the physical properties of the RF discharge on the spatial and other

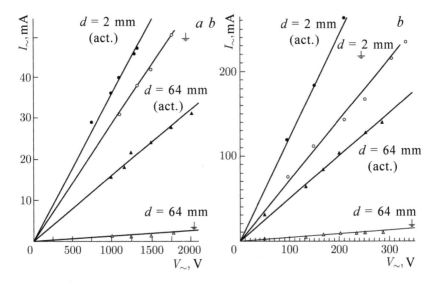

Fig. 1.47. CVC of the discharge gap (without discharge), depending on the magnitude of the interelectrode distance d. a) $f = 1.2$ MHz; b) $f = 13.56$ MHz.

characteristics of the discharge gap, let us consider the simplest form of the latter – plane electrodes of diameter D, spaced at a distance d.

First, we note how the properties of the RF discharge are related to the interelectrode distance d.

Experimental studies show that in RFCD there is a significant difference in the currents flowing in the electrical circuits of the active (AE) and grounded (GE) electrodes. This means that not all of the current from the AE reaches the GE. This situation, judging by the literature, has been little studied.

Work [145] experimentally studied the dependence of the current fraction on the AE reaching the GE from the distance d between the electrodes.

To compare the degree of the ability to short-circuit the bias current of the RF field at frequencies of 1.2 and 13.56 MHz on a grounded electrode for a methodical purpose in the absence of a discharge, an I–V characteristic was obtained in electric circuits of AE and GE for various d (Fig. 1.47). It can be seen that for small distances ($d \approx 2$ mm), the current closure at 1.2 MHz is higher than for the 13.56 MHz frequency, and for sufficiently large distances ($d \approx 64$ mm) the degree of closure is much lower and practically independent of the field frequency.

The shape of the current–voltage characteristics of the RFCD also depends on the distance d. The families of the I–V characteristics,

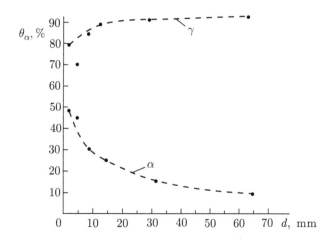

Fig. 1.48. Dependence of the fraction θ_α (%) of the total current $I_\sim$, which is closed on the grounded electrode, on the interelectrode distance d. RFCD, air, $p = 30$ Torr, $f = 1.2$ MHz; α – before the jump $\alpha \rightarrow \gamma$; γ is the first point after the jump $\alpha \rightarrow \gamma$.

where d is the parameter, were presented in Fig. 1.18 and were used to determine the dependence of the fraction of the current of the AE that reached the GE from the interelectrode distance d. In this case, the α-branches of the current–voltage characteristics of AE and GE were compared for various d, using for definiteness the value of the current $I_\sim$ corresponding to the beginning of $\alpha \rightarrow \gamma$-transition.

The obtained dependence of the current fraction of the active electrode θ_α (%), which closes on the grounded electrode, on the distance d is shown in Fig. 1.48.

It is seen that the quantity θ_α (%) decreases monotonically with increasing interelectrode distance d. An analogous value of θ_γ and a dependence of $\theta_\gamma(d)$ for the γ-branch of the current–voltage characteristic of the RFCD under consideration are also shown in Fig. 1.48. For comparison, the first points of the γ-branch of the current–voltage characteristic for AE and GE were used.

Let us pay attention to the observed opposite behaviour of the dependences θ_α (d) and θ_γ (d) with increasing d: on the α-branch the quantity θ_α decreases, and on the γ-branch the value θ_γ increases.

Such a behaviour of the currents in the AE and GE chains with increasing d is explained by the fact that: 1) only the factor acting on the α-branch increases the impedance of the discharge gap in comparison with the impedance from the AE to the surrounding space, namely, the reduction of the interelectrode capacitance C_d; 2) in the case of the γ-branch, there are a number of factors that substantially reduce the impedance of the discharge. Indeed, as can be seen from Figs. 1.18 a and 1.18 b, with increasing d, the voltage $V_\sim$ and the current $I_\sim$ increase at which the jump occurs, and the γ-branch begins with large values of the currents $I_\sim$, while the capacitances of the near-electrode layers C_{s1} and C_{s2} increase and the conductivity of the gas-discharge plasma, as if bringing the electrodes together.

It should be noted that the nature of the manifestation of the described current-reconnection effect in AE and GE circuits after the jump $\alpha \rightarrow \gamma$ depends strongly on the nature of the gas.

Depending on the mode of the RFCD, its spatial structure can be different (α-discharge, γ-discharge). As shown in Ref. [67], the detailed structure of the γ-type RFCD completely coincides with the structure of DCGD [151]. Moreover, when the electrodes approach each other, all spatial regions disappear – analogues of the DCGD regions, with the exception of the near-electrode regions. Thus, the magnitude of the interelectrode distance d determines, fully or in part, the spatial structure of RFCD at its prescribed external parameters (pressure and gas type, amplitude $V_\sim$ and frequency f of RF voltage, diameter of tube D).

The vertical sections of the I–V characteristic (Figs. 1.18, 1.19) are sections of the 'normal' RFCD regime, similar in its properties to the corresponding DCGD mode. From this, in particular, it can be seen that with increasing d, the 'normal' mode section is reduced along the current interval and shifts toward increasing the RF voltage $V_\sim$ supporting the discharge.

As will be shown later, analysis of the family of the I–V characteristics with different d gives a significant amount of information on the RFCD characteristics [75]. It is established that for small d, a symmetrical RFCD is a two-electrode discharge, and for large d, two autonomous single-electrode discharges are possible.

Indeed, one of the interesting features of RFCD is the possibility of its implementation in the presence of a single electrode with the closure of the discharge circuit to the surrounding discharge gap by the 'ground'.

One-electrode RF discharges at atmospheric pressure have been known for a long time – corona and torch discharges [1].

The single-electrode low- and medium-pressure RFCD in the discharge tube, which is actually entirely due to the edge effect of the RF field, has not been practically investigated.

Meanwhile, a detailed experimental study of the I–V characteristics of a single-electrode RFCD (RFCD-1) certainly contributes to a deeper understanding of the physical meaning of the I–V characteristic of an ordinary RFCD with two electrodes (RFCD-2).

In connection with what was said above, in a number of experiments the I–V characteristics were obtained in the same physical conditions in one discharge tube alternately for RFCD-2 and RFCD-1. And in the latter case, the grounded electrode simply disconnected from the laboratory 'ground'.

The CVCs of the RFCD-2 and RFCD-1 obtained in the course of the experiments in AE and GE circuits using short discharge gaps ($d = 5$ mm) are shown in Fig. 1.24.

The current–voltage characteristic in the AE RFCD-2 and RFCD-1 circuits for a sufficiently long length $d = 32$ mm is shown in Fig. 1.25.

Naturally, in these experiments the RFCD-2 and RFCD-1 current–voltage characteristics in the AE circuit were compared.

The CVC of the the RFCD-2 in the GE circuit is presented for completeness of the picture, and the CVC of the RFCD-1 in the GE circuit (a freely 'hanging' electrode) is presented only to illustrate the magnitude of the bias current that hits this electrode.

Let's analyze the received CVCd for RFCD-1 and RFCD-2.

First of all, note that the CVCs of both these discharges have two branches (lower and upper), separated by a jump.

The lower branch of both discharges will be called the 'α-branch'. The upper branch of RFCD-2, by tradition, is called the 'γ-branch', and for the upper branch of RFCD-1 we will not introduce a special name.

Note that if the α-branches of both discharges are almost identical, then the upper branches of RFCD-1 and RFCD-2 differ in principle. In particular, on the γ-branch of RFCD-2, the phase shift between the RF current $I_\sim$ and the voltage $V_\sim$ is much less than $\pi/2$, while on the entire RFCD-1 current–voltage curve this phase shift $\varphi \approx \pi/2$, as well as on the α-branch of RFCD-2.

Since the CVC of the RFCD-2 is discussed in detail above, we will mainly concentrate on the analysis of the RFCD-1 current–voltage characteristics.

In contrast to the γ-branch of the RFCD-2 CVC, identical in shape and significant contribution to the total current of the conduction current of the branch of the 'normal' DCGD mode, the upper branch of the RFCD-1 current–voltage characteristic behaves similarly to the α-branch of the capacitive current, $I_\sim = \omega C V_\sim$, where C is the effective capacitance between the AE and the 'ground' of the space surrounding the discharge tube.

Note that for all the CVCs of the RFCD-1, shown in Figs. 1.24 and 1.25, the angle of inclination of the straight line of the upper branch is greater than the lower one. This means that after the jump on the RFCD-1 current–voltage characteristic, the effective capacitance C increases.

The jump at the RFCD-1 current–voltage characteristic is caused by ignition of the discharge in the discharge tube and the appearance of a plasma that causes the formation of space charge layers near the inner surfaces of the tube. The appearance of a spatial charge layer should lead to an increase in the said effective capacitance C.

Let us also pay attention to the fact that after the jump in the CVC of the RFCD-2 the RF voltage takes the value $V_{\sim n}$, close to the tabulated value of the 'normal' cathode dip V_{cd} in the DCGD [151], the closer the interelectrode distance d is.

After the jump at the RFCD-1 current–voltage characteristic, the RF voltage assumes a value $V'_\sim$, much higher than the value V_{cd} in the DCGD under similar physical conditions.

What has been said above means that the mechanisms for maintaining the RFCD-1 and RFCD-2 after the jump on the I–V characteristic differ in principle.

In addition, as seen in Figs. 1.24 and 1.25, with increasing distance d between the electrodes, the α-branch of RFCD-2 approaches the α-branch of RFCD-1.

This can be interpreted in such a way that before the jump on the CVC the RFCD-2 behaves like an RFCD-1, that is, also in the presence of two electrodes, for sufficiently large d, the RFCD-1 is realized before the jump. After a jump on the current–voltage characteristic, the presence of a sufficiently dense plasma in the discharge gap ensures an efficient short-circuit of the current of the AE circuit to the GE and a transition to the real RFCD-2 occurs.

Using as an example the I–V characteristic in Fig. 1.24, we estimate the effective capacitance C between the RFCD-1 electrode and the 'earth' of the surrounding space after a jump on the current–voltage characteristic for a current $I_\sim = 60$ mA:

$$C = \frac{I_\sim}{\omega V_\sim} = \frac{60 \cdot 10^{-3}\,(\text{A})}{2\pi \cdot 1.2 \cdot 10^6\,(\text{Hz}) \cdot 1500\,(\text{V})} = 5.3 \cdot 10^{-12} \quad \Phi = 5.3\,\text{pF}.$$

As a result, we get a plausible value for C.

On the basis of what has been said above, it can be asserted that for sufficiently large distances d and relatively low RF voltages $V_{\sim}$, even a symmetrical RFCD with both active electrodes can be two autonomous single-electrode discharges, and not a single two-electrode discharge.

The most characteristic configuration of the widely used RFCD is an asymmetric discharge (ARFCD) with electrodes of different areas, one of which is grounded.

Note that when grounding one electrode, even a discharge with electrodes of equal area will be almost asymmetric. Indeed, since the actual discharge gap, due to parasitic capacitances, is associated with the surrounding 'ground', an 'effective' grounded electrode will have an area larger than the area of the real electrode.

The physical properties of ARFCD are described in some detail in the monograph [6]. Therefore, here we only briefly mention some of its features: 1) the existence of two discharge modes – V_0-mode (with a blocking capacitance in the discharge circuit) and I_0-mode (the discharge circuit is closed in direct current); 2) the possibility of the occurrence in the NESCL of an active electrode of an instantaneous voltage of the order of twice the amplitude RF voltage $V(t) \leq 2V_{\sim}$; 3) anharmonicity of RF discharge current; 4) electron escape from the discharge gap in the I_0-mode only through the active electrode of a small area.

Since the anharmonicity of the current in the ARFCD is considered earlier, we shall not dwell on this.

If the active electrode is not short-circuited to the ground (via the RF choke), then using or without using the blocking capacitance C_{bl} in the discharge circuit, one can create two modes in the ARFCD: V_0-mode and I_0-mode.

In the case of the V_0-mode, the discharge circuit is disconnected by direct current, and self-bias in the discharge arises along a constant electric potential, that is, a quasi-stationary potential difference equal to the difference in the jumps of the constant potential in the NESCL of both electrodes.

When maintaining the ARFCD in the I_0-mode (without a blocking capacitance C_{bl} or shorting the active electrode to ground through the blocking RF choke), a constant component of the discharge current appears in the interelectrode gap (the 'battery' effect).

We note the following (important for practical applications) peculiarity of the voltage behaviour in the NESCL of the active electrode in the ARFCD. If, even in a symmetrical discharge, the

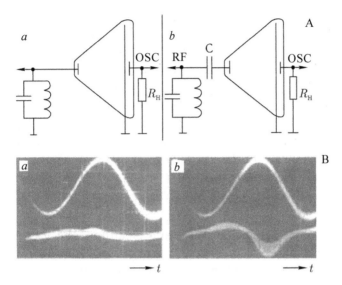

Fig. 1.49. Investigation of electron fluxes onto a grounded electrode of a large area in the ARFCD. A. Experimental scheme: a) I_0-mode (without blocking capacitance C_{bl}); b) V_0-mode (with a blocking capacitance C_{bl} = 3300 pF). B. Oscillograms of signals I_k (t) from the collector of the energy analyzer: a) I_0-mode; b) V_0-mode.

instantaneous potential difference in the layer mentioned above can be simply greater than the amplitude value $V_\sim$, then in ARFCD, in the case of strong asymmetry, there can be a significant increase in the instantaneous voltage in the layer $V(t) \leq 2V_\sim$.

Finally, let us pay attention to the specifics of the functioning of ARFCD electrodes in the I_0-mode established in the theoretical work [78], when electrons from the discharge gap leave only through the active electrode of small area. In this case, a stationary potential barrier, which holds electrons, is installed in front of a large-area grounded electrode.

The above-mentioned character of the escape of electrons from the discharge gap ARFCD was experimentally confirmed in [145]. The measurement scheme used is shown in Fig. 1.49 *a*. On the obtained oscillograms of the charge flows to the grounded ARFCD electrodes in the V_0- and I_0-modes it is seen that in the case of the V_0-mode there is a current pulse of electrons to the electrode in the positive half-period of the field, and in the I_0-mode electrons on the ground electrode do not fall (Fig. 1.49 *a, b*).

In the case of the need to use near-electrode plasma, a transverse RFCD with a small interelectrode distance *d* is used.

As is known [67, 145], high-energy near-electrode electron beams supporting a discharge appear in the γ-type RFCD. With this in mind, when using the transverse RFCD, a distance d commensurate with the relaxation length of the electron beams in terms of energy is selected.

At low-pressure RFCD ($p < 10^{-1}$ Torr), the near-electrode electron beams can excite beam-plasma instabilities (BPI) of the Cherenkov type [186]. Therefore, in order to establish the possibility of the appearance of the latter, it is necessary to compare the distance d with the characteristic length L of their swing [155]

$$L \approx \frac{v_{eb}}{\delta},$$

where v_{eb} is the velocity of the beam electrons, and δ is the growth rate of the instability. When the necessary conditions for the excitation of the BPI and the additional condition $L < d$ are fulfilled, the instabilities are excited, the beams effectively decay, and a considerable heating of the plasma electrons by the microwave fields of the instabilities is observed.

The length of the discharge gap d determines the frequency of the longitudinal resonance ω_r arising in the low-pressure RFCD at operating frequency ω below the electronic plasma frequency ω_{0e}, when the gas-discharge plasma behaves as an inductance and resonates with the capacitance of the near-electrode layers of space charge. The resonant frequency is given by the expression [187]:

$$\omega_r \approx \omega_{0e} \left(\frac{d_s}{d_s + d_p} \right)^{1/2},$$

where d_s is the total thickness of the near-electrode layers, d_p is the length of the gas-discharge plasma, $d = d_s + d_p$. Because of this resonance, high, oppositely directed voltage drops on the layers and plasma arise in the discharge [188].

We now note some points related to the dependence of the RFCD properties on the diameter of the electrodes D. Additional information on this issue is also reported below in connection with the consideration of the role of the discharge gap configuration in the RFCD mechanism.

First we point out the universal circumstance: it is easier to maintain the stationarity of the balance of any particles with a large value of the ratio (volume / area) for the region of interest, which is achieved for commensurable values of the diameter D and the length d of the discharge gap.

From particular moments we can mention, for example, the situation with the excitation of a collisional–dissipative instability with an increment δ [6]:

$$\delta = \omega_{0e} \left(\frac{n_{eb}}{n_e} \cdot \frac{\omega_{0e}}{v_{en}} \right)^{1/2},$$

where n_{eb}, n_e are the electron beam and plasma electron densities, respectively; v_{en} is the collision frequency of electrons with neutral particles. From this it is clear that the increment δ is greater the higher the density of the beam n_{eb}. At a fixed power deposited in the RFCD, it is obvious that the value of n_{eb} will be larger with a smaller diameter of the electrode D.

At present, more and more developers are attracted by the problem of creating plasma equipment with a large area of processed surfaces and using RF frequencies higher than the usual industrial frequency of 13.56 MHz. Large-area plasma sources were required when manufacturers of microelectronics products increased the diameter of silicone substrates to 0.3 m and also required the processing of large (1 m × 1 m) glass plates for active matrices of low-crystalline flat panels of computers and television screens. In this case, a problem of the homogeneity of the processing process arose, in particular, because of the occurrence of standing waves at the surface of extended electrodes. In the theoretical paper [146], the condition for creating a uniform RF discharge without appreciable standing waves

$$\lambda_0 \gg 2.6 \left(\frac{l}{s} \right)^{1/2} R,$$

where λ_0 is the wavelength of the RF field in free space, $2l$ is the distance between the electrodes, s is the thickness of the near-electrode layer, and R is the radius of the electrode.

In real discharge gaps of the RFCD there are always edge effects. Thus, in a discharge of planar geometry in the near-electrode regions, in addition to the longitudinal electric field, there is a radial component of the RF field emerging from the discharge gap. It was shown in [189] that this field component initiates the appearance of electron beams in the space charge layer near the dielectric wall of electron beams directed radially to the axis of the discharge tube. These wall beams return some of the energy of the RF field lost as a result of edge effects to the discharge gap, and in addition to the near-electrode beams enrich the gas-discharge plasma with higher-energy electrons. At the RFCD with small electrode spacings d, practically the entire surface of the dielectric cavity around the discharge gap becomes a source of radial wall electron beams.

For completeness, we note that the minimal edge effects are characterized by a fairly extended RFCD of coaxial geometry, whose RF field is almost completely localized between the electrodes. In the experimental studies of RFCD, as a rule, and in its practical applications, one of the electrodes is always grounded. This leads to the fact that even with completely identical electrodes, the RFCD is somewhat asymmetric with a somewhat uncertain configuration.

As a result of what was said earlier, the current–voltage characteristics measured in the electrical circuits of the active and grounded electrodes differ significantly (Fig. 1.19 a, b). Depending on the physical conditions of the experiment, up to 90% of the discharge current in the active electrode circuit can go to the ground through the parasitic capacitors, and only the rest of the current falls on the ground electrode, being the true discharge current of the RFCD.

Of particular interest is the RFCD with external electrodes, i.e. electrodes covered with a dielectric. Strictly speaking, in such a discharge, the electrode is the inner surface of the dielectric, which borders on the space charge layer, to the outer surface of which the

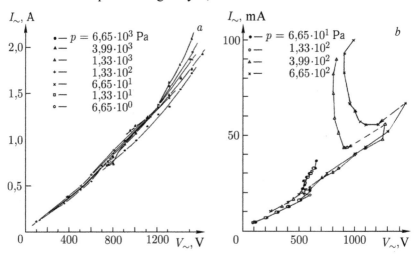

Fig. 1.50. Dependence of the current–voltage characteristics of the RFCD with external electrodes on the dielectric type. a) Xe, $f = 1.2$ MHz, $d = 2 \cdot 10^{-2}$ m, dielectric – quartz. b) the air, $f = 1.2$ MHz, $d = 3 \cdot 10^{-2}$ m, and the dielectric is $BaTiO_3$.

metal electrode with RF voltage is closely brought. The advantage of such a discharge is the absence of contact of the metal electrodes with the plasma, leading to the sputtering of the metal, as well as the chemical modification of the electrode, especially in the case of an aggressive plasma-forming medium.

In the case of traditionally used dielectrics (glass, quartz) with a dielectric constant $\varepsilon \approx 5$, the impedance of the insulating layer exceeds the impedance of the discharge gap. Therefore, the current–voltage characteristic with external electrodes is practically independent of the physical conditions in the discharge (in particular, of the gas pressure) and represents a monotonically increasing curve (Fig. 1.50 *a*). The character of the current–voltage characteristics changes significantly when using the ferroelectric TiBa ($\varepsilon \approx 2000$) as an insulating material, when the drop in the RF voltage at the impedance of the insulation decreases sharply (Fig. 1.50 b). A feature of this I–V characteristic is the presence of a stable falling section of the characteristic noted earlier in the study of RFCD with external electrodes in the work [190].

The magnitude of the quasi-stationary potential difference between the RFCD electrodes equal to the value cut off on the stress axis of a straight line constructed according to the method described in [145] was measured with the help of the CVC family (Fig. 1.19 b).

It has been experimentally established that the characteristics of RFCD essentially depend on the shape of the electrodes. Electrodes

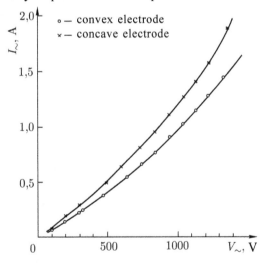

Fig. 1.51. The dependence of the current–voltage characteristics of the RFCD on the sign of the curvature of the electrodes of radius $R = 3$ cm. He, $p = 39$ Torr, $f = 1.2$ MHz.

with positive and negative radii of curvature were used, that is, convex and concave. The dependence of the form of the CVC of the RFCD on the sign of the radius of curvature of the electrodes is shown in Fig. 1.51.

It can be seen that the CVC of the concave electrode passes above it for a discharge with convex electrodes. During the experiments, it was found that the near-electrode beams are scattered to the walls of the discharge tube in RFCD with convex electrodes and their focusing to the axis of the discharge tube with concave electrodes took place. Obviously, in the second case, the range of the beam runs (and, correspondingly, ionization and excitation of the gas) in the discharge is much larger than in the first. This explains the location of the corresponding I–V characteristic in Fig. 1.51.

By using different electrode shapes (flat, 'cup', ring, etc.), different configurations of electron beams were implemented in RFCD.

A strongly asymmetrical RFCD with a small area of one of the electrodes ('tip–plane' type) is used at high pressures to produce a gas breakdown by means of a large field strength at a small electrode. One-electrode RFCD is also an example of a sharply asymmetric discharge, since the discharge current is closed by a bias current to a second electrode of a large area ' surrounding the 'ground'.

A specific type of RFCD is a discharge with electrodes from a metal mesh. In the course of experiments using a medium-power RF generator in such a discharge, it was not possible to obtain a transition from the α-discharge to the γ-discharge. Apparently, this is

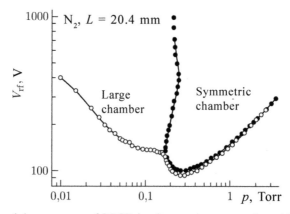

Fig. 1.52. Breakdown curves of RFCD in nitrogen in symmetric and large chambers with a discharge gap $d = 20.4$ mm [191].

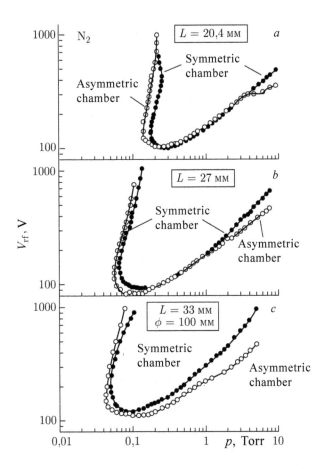

Fig. 1.53. Curves of breakdown of RFCD in nitrogen in symmetric and asymmetric chambers. a) $d = 20.4$ mm, diameter of electrodes $D = 143$ mm, b) $d = 27$ mm, diameter of the electrodes $D = 143$ mm, c) $d = 33$ mm, diameter of the electrodes $D = 100$ mm [191].

due to the small area of the electrodes and the low average density of emission processes that are not capable of supporting the γ-discharge.

Above are given data on the dependence of the physical properties of RFCD on the configuration of discharge gaps in conventional discharge tubes intended for academic research. Naturally, it is of considerable interest to study the characteristics of RFCD under conditions of a real technological discharge chamber.

In [191], the influence of the geometry of the discharge chamber on the ignition process and the characteristics of the low-pressure RFCD was studied experimentally and numerically. The discharge was excited at a frequency of 13.56 MHz in chambers with three different geometries: parallel plates surrounded by a dielectric cylinder ('symmetrical chamber'); parallel plates surrounded by a grounded metal cylinder ('asymmetrical chamber') and parallel plates inside a much larger grounded metal chamber ('large chamber').

The obtained breakdown curves for a symmetric chamber have a region of ambiguous dependence $V_B(p)$ at low pressures. In the geometry of a large chamber, the ambiguity region is not observed (Fig. 1.52).

For an asymmetric chamber, the breakdown curves are shifted to low-pressure and low RF voltage regions, but the ambiguity region is still present, at higher pressures the breakdown voltage $V_{\sim B}$ is much lower than for the symmetric geometry (Fig. 1.53).

As can be seen from the above data, the geometry of the discharge chamber greatly affects the characteristics of RFCD. More detailed information on this account and an explanation of the mechanisms of this influence are given in [191].

In connection with the question of the role of the configuration of the discharge gap of the RFCD, mention should be made of discharges used in laser technology. These are varieties of the transverse RFCD when the RF field is directed along the minimum size of the electrode configuration: 1) semicylindrical electrodes along the length of the cylindrical discharge tube; 2) waveguide (capillary) electrode system [192]; 3) slotted electrodes [6].

Based on the above material, it can be concluded that the physical properties of RF gas-discharge systems (the spatial structure of the RF discharge, the set of physical processes and phenomena that make up the RF discharge mechanism, the parameters of the gas discharge plasma) can be effectively controlled not only by changing the external discharge parameters (pressure and kind of gas, amplitude and frequency of RF voltage, electrode material), but also by using different configurations of the discharge gap and the shape of the electrodes.

1.10. Dynamics of physical processes in the RFCD

In a stationary RFCD there is a dynamics of physical processes within each period of the RF field.

One should immediately note the fundamental difference in behaviour of α- and γ-discharges in this respect.

In the α-discharge, the processes in the NESCL practically do not affect the time course of the processes in the discharge gap. In the γ-discharge, on the contrary, it is the time dependence of the parameters of the NESCL that determines the nature of the time dependence of the physical conditions in the RFCD.

In accordance with the well-known physical mechanism of the α-discharge, the plasma of the latter is maintained by the RF field present in the discharge gap. The intensity of the processes of ionization and excitation of the gas is determined by the local value of the RF field strength $E_\sim(x)$. The depth of time modulation with the frequency of the field of plasma parameters in the volume of the discharge is determined by the characteristics of the physical conditions in the discharge (the pressure and genus of the gas, the strength and frequency of the RF field, the geometry of the discharge gap).

In the physical mechanism of the RFCD of the γ-type, as is well known, an important role is played by the developed NESCLs, in which, due to the phenomenon of RF detection, significant quasi-stationary electric fields arise [28]. These near-electrode fields form high-energy pulsed electron beams injected from the electrodes into the plasma.

It is obvious that in symmetrical and asymmetric RFCDs the course of the processes in the discharge gap will differ substantially due to the appearance of a significant anharmonicity of the discharge current due to non-linear effects in the second case.

The parameters of the above-mentioned NESCL (the jump in the quasi-stationary electric potential in the layer U_s, the layer thickness d_s) are modulated by the external RF voltage. In general, during one period of the high-frequency field, U_s can take values from 0 to $U_{s\,max}$, and the layer thickness d_s varies from 0 to $d_{s\,max}$.

It should be noted, however, that as the frequency f and the RF field voltage $V_\sim$ increase, according to the experimental data [145], the layer length d_s tends to a constant value, and the modulation depth of the quasi-stationary voltage U_s decreases substantially. In addition, there was also a large decrease in the amplitude of the luminescence oscillations in the region of the layer boundary at the frequency of the RF field, which will be discussed later.

The quasi-stationary spatial structure of the integral luminescence of the y-type RFCD was presented earlier in Fig. 1.33.

Investigations of the dynamics of the spatial structure of the luminescence in the near-electrode region of RFCD during one period of the RF field were also carried out using an experimental scheme whose optical path included an optical monochromator and a C7-13 strobe oscilloscope used as a gating element (Fig. 1.54).

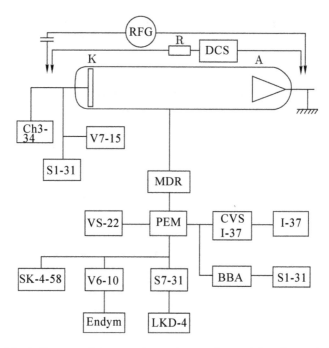

Fig. 1.54. Block diagram of the experimental setup for studying the spatial-temporal distribution of the luminescence intensity in the discharge gap of the RFCD. RFG- RF generator Bryg; DCS – DC source; ChZ-34 – frequency meter; V7-15 – voltmeter; S1-31 - the oscilloscope; MDR – monochromator MDR-12; VC-22 – a source of a constant voltage; PEM – photomultiplier FEU-79; CVS I-37 – direct current amplifier; I-37 – recorder; SC-4-58 – spectrum analyzer; V6-10 – selective voltmeter; S7-13 – the oscilloscope; BBA – the broad-band amplifier; Endym – chart recorder; LKD-4 – recorder.

In this case, the intensities of a number of spectral lines were registered in certain phases φ of the period of the RF field. The spatial distribution of the intensity of each line was obtained by moving on the electric platform of the discharge tube in front of the input of the monochromator.

Typical 'phase' curves obtained for the intensity distribution of the lines $I_\lambda(\varphi, x)$ for various RF voltages $V_\sim$ and gas pressures p are shown in Fig. 1.55.

As can be seen from the obtained distributions $I_\lambda(\varphi, x)$, the qualitative form of the luminescence structure remains unchanged during the full period of the RF field. From this we can conclude that in the investigated RFCD modes the plasma did not touch the electrode surface at any time.

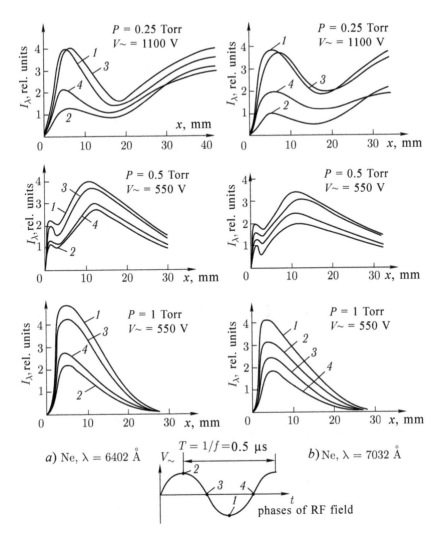

Fig. 1.55. 'Phase' curves for the space distribution of the intensity of the spectral lines Ne *I* = 6402 Å and Ne *I* = 7032Å for various modes of RFCD.

Thus, the currently used simplified model of the behaviour of the NESCL of the RFCD [6], in which the thickness of the layer varies from the maximum value $d_{s\,max}$ to the total disappearance (layer collapse), is not always true. At the same time, the characteristics of the NESCL are often radically changed during the period of the field *T* and, accordingly, the parameters of the near-electrode electron beams (n_{eb} density and energy ε_{eb}) are deeply modulated in time.

It follows from general physical considerations and experimental data [169] that there exists an interval of 'active phases' of the period of the RF field T in the neighborhood of the phase with the amplitude value of the RF voltage $V_\sim$ in the negative half-cycle when the quantities n_{eb} and ε_{eb} are maximal. It follows that the intensity of ionization and excitation of the gas by electron beams varies with time.

An important circumstance affecting the dynamics of physical processes in low-pressure RFCDs is the appearance, under suitable conditions, of beam-plasma instabilities (BPI), considered below. The electric fields of the latter, generated in the microwave range, arise in a certain phase of the RF field period. The local heating of electrons due to them reaches effective temperatures of the order of hundreds of eV, which should be accompanied by a significant increase in the electrical conductivity of the plasma. Then the phase of the RF field is reached, when, due to a number of reasons, BPIs break and the additional heating of the electrons and the effective attenuation of the electron beam are interrupted.

The above phenomena affect the dynamics of physical processes in the discharge. According to the foregoing, all the processes and phenomena that determine the dynamics of physical conditions in RFCD are due to two factors: 1) RF field and 2) near-electrode pulsed electron beams.

As is known [1], in the RF field the processes of interaction of electrons with gas occur with a doubled frequency of the field $2f$. The interaction of the near-electrode electron beam with plasma particles is characterized by the frequency of the field f [67]. Therefore, the spatial distributions of the amplitude of the luminescence oscillations at frequencies f and $2f$ given earlier qualitatively show sections of the discharge gap with the corresponding dynamics of physical processes (Figs. 1.41–1.43).

Concluding a brief discussion of the dynamics of physical processes during the RF field, it should be noted that the nature of the change in physical processes over time in RFCD depends, generally speaking, on all discharge parameters. For example, the range of changes in the parameters of the NESCL (U_s, d_s) depends on the amplitude $V_\sim$ and the frequency f of the RF voltage; the excitation conditions of the BPI depend in particular on the interelectrode distance d and the diameter of the electrodes D

1.11. The frequency of the RF field – a fundamental parameter of the RFCD

The frequency of the RF field f is a fundamental parameter of the RFCD.

It should be noted that the effect of frequency on the physical properties and the characteristics of the discharge under study have not yet been investigated experimentally enough.

The presence of an additional parameter f in the RF discharge, compared to a DC discharge, leads to a large difference in the physical content of these discharges.

The fundamental difference between the properties of RFD and DCD is due, first of all, to the physical nature of their total discharge current: in the DCD there is only a pure conduction current, and in RFD – the sum of the current of conductivity and bias current. The latter, as is well known, depends strongly on the frequency of the field.

According to the foregoing, the RF discharge occupies a certain frequency range of the electromagnetic field, in which the discharge can be maintained by both potential and vortex electric fields, depending on the frequency of the field.

In contrast to the situation with the DCD, there is a problem of introducing RF power at different frequencies into the discharge gap, solved with the help of special radio devices that match the output impedance of the RF generator with the impedance of the RFD.

Recently, in connection with the increased demands of high technologies, two- and three-frequency RF plasma chemical technological reactors began to be used in which the parameters of plasma and near-electrode layers of space charge can be independently controlled.

The frequency of the field is an important parameter that determines the nature of the motion of charged particles in the RF field, the process of energy collection by electrons, while the ratio of the frequencies of the field f and collisions of electrons with neutral particles v_{en} is of significant significance.

With increasing frequency f, due to the vibrational motion of charged particles, their escape from the discharge gap is reduced and the charge density n_e increases.

We have already noted the frequency dependence of the breakdown voltage V_B in a gas when the minimum of the $V_B(pd)$

curve was observed under the condition $\omega \sim \nu_{en}$, where ω is the angular frequency of the field.

The interaction of the RF field with the near-electrode plasma under the RFCD conditions leads to the appearance of the reducing effect, which also has a frequency-dependent character.

It is obvious that the nature of the change in time of all physical processes in RFCD is determined by the frequency of the RF field.

In an α-discharge, the RF field with a doubled frequency $2f$ modulates the energy of the plasma electrons, and, consequently, the rate of elementary processes in the plasma.

Significantly more diverse are the manifestations of the influence of the field frequency on the dynamics of the RFCD properties at high RF voltages, when the γ-discharge regime is realized. Here, first of all, two effects in the near-electrode regions should be mentioned: complete or partial collapse of the NESCL and the formation of pulsed electron beams. In particular, this refers to short discharge gaps where the plasma parameters, mean energy $\bar{\varepsilon}_e$ and electron density n_e, can be substantially modulated in time.

At higher frequencies f, the ion motion in the NESCL becomes more and more steady, and the properties of the near-electrode layer approach the properties of the cathode drop region of the DCGD. Under these conditions, the process of the entry of plasma electrons by diffusion to the electrodes becomes difficult because of the small time intervals favourable for this, which will be considered below.

The known stochastic mechanism for heating plasma electrons by the oscillating boundary of the NESCL [56] also depends on the frequency of the field.

As is widely known [6, 145], the plasma charge density strongly depends on the frequency $n_e \sim f^2$. In connection with this, as f increases, the density of ion fluxes to the electrode increases and, correspondingly, the density of the emitted electrons n_{e0} forming the near-electrode electron beams also increases.

It will be shown below that in the low-frequency range f, as a result of the frequency dependence of the nature of the ion motion in the neighbourhood of the electrode surface, the quasi-stationary jump in the electric potential U_s in the NESCL will depend substantially on the field frequency.

From the foregoing it follows that the beams in RFCD are modulated at a frequency of f, so that they can cause the excitation of plasma instability at the frequency of the RF field. Here it should be noted that as the field frequency increases, the beam lifetime

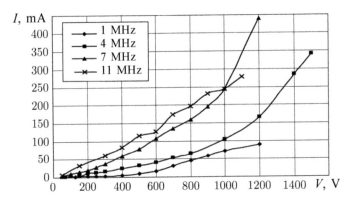

Fig. 1.56. CVCs of RFCD with external electrodes at different frequencies of the RF field (tube diameter 80 mm). Gas – neon, $p = 0.5$ Torr.

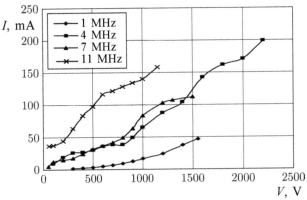

Fig. 1.57. CVCs of RFCD with external electrodes at different frequencies of the RF field (tube diameter 40 mm). Gas – neon, $p = 0.5$ Torr.

$\tau_{eb} \sim f^{-1}$ decreases and, accordingly, the duration of all the beam effects produced in the plasma also decreases.

The possibility of excitation of beam-plasma instabilities in the plasma of the RFCD, considered below, is undoubtedly connected with the frequency dependence of the excitation condition of BPIs. This is due to the dependence of the plasma and beam parameters on f and, first of all, the duration of the latter.

The dependence of both physical factors supporting RFCD – RF field and the near-electrode beams on the frequency of the field f finds a peculiar reflection in the behaviour of such a complex characteristic of the discharge as its I–V characteristic.

In [193] it was experimentally found that in the CVC family obtained in the same discharge tube for different frequencies of the

RF field there exists an I–V characteristic at a certain frequency f with the largest value of the discharge current $I_\sim$ for a fixed RF voltage $V_\sim$. Thus, for the given experimental conditions, there is an optimal frequency f to support the RFCD.

Characteristic families of the I–V characteristics obtained upon excitation of the RFCD in discharge tubes with one gas filling, but with different diameters D, are shown in Figs. 1.56 and 1.57.

We note the experimentally established in this work specific nature of the dependence of the CVC of the RFCD on the frequency of the RF field, f, which in turn depends on the diameter D of the discharge tube.

As is known [67], with an increase in the field frequency, the charge density of the plasma n_e increases substantially at a constant amplitude of the RF voltage $V_\sim$, which is fairly well studied in the α-discharge. Therefore, we should expect a larger value of the discharge current $I_\sim$ for the I–V characteristic corresponding to a higher frequency f at moderate values of the RF voltage $V_\sim$. Such a picture is observed with respect to the I–V characteristic presented in Figs. 1.56, 1.57, where the diameters of the discharge tubes differed by a factor of two.

At the same time, as can be seen from the current–voltage characteristics, in the region of increased voltages $V_\sim$ corresponding to the γ-discharge, the current–voltage characteristic with a lower frequency f can 'overtake' the I–V characteristic corresponding to a higher frequency. Thus, in Fig. 1.56 (for the case of a discharge tube of diameter $D = 0.08$ m) for $V_\sim > 1000$ V, the current–voltage characteristic at a frequency of 7 MHz is higher than the I–V characteristic corresponding to 11 MHz.

A similar situation is observed in Fig. 1.57 for a tube of half the diameter ($D = 0.04$ m), when the I–V characteristic at 4 MHz becomes 'active', 'overtaking' the I–V characteristics at a frequency of 7 MHz.

Recall that in the case of an α-discharge the main factor supporting the discharge is the RF field and in the γ-discharge it is the near-electrode electron beams. Moreover, the fact that electron beams are pulsed with a pulse duration of the order of $1/f$ is of considerable importance here. Thus, with increasing frequency field, the lifetime of the beams (and, consequently, the duration of the ionization processes produced by them and the excitation of the gas) decreases. On the other hand, as the frequency f increases, the escape of charges from the discharge gap slows down, and as the condition $\omega \approx v_{en}$ is approached (v_{en} is the collision frequency of electrons with atoms), the energy contribution of the field to the plasma increases. This increases the charge density ne of the plasma.

Thus, the mechanism of the frequency dependence of the charge density n_e (and hence the CVC) in the discharge under study is quite complex. Naturally, the balance of charged particles in the discharge must depend substantially on the diameter of the discharge tube, which affects the loss of charges in the radial direction. The latter is confirmed by the differences in the frequency dependences detected by the CVC curves in different diameter tubes (Figs. 1.56, 1.57).

Concerning the case of RFCD with external electrodes, it should be noted that as the field frequency increases, the RF voltage decreases at the capacitive impedance of the dielectric layer that insulates the electrodes.

Finally, from a purely technical point of view, it should be noted that with the increase in the frequency of the RF field, the role of all parasitic capacitances and leakage of RF power to the ground increases and electromagnetic pickups on diagnostic equipment are amplified.

2

Physical properties of near-electrode layers of space charge of RF capacitive discharge

The near-electrode layers of space charge play an important role in the physical mechanism of RFCD. They strongly affect the electrical characteristics of the discharge as a whole and its plasma parameters. The physical processes in the NESCL (near-electrode space charge layer) directly affect the type and rate of reactions on the surface of the discharge electrodes and substrates in technological plasma-chemical reactors.

2.1. The specificity of the NESCL of RFCD in comparison with DC discharge

In the RFCD there is no such fundamental difference in the functions of the electrodes as the presence of an anode and a cathode in the DCD (DC discharge). The electrodes of the RFCD are functionally the same, and in a symmetrical discharge without the earthing of one of the electrodes they are completely identical. Analogous to the DCD, it can be said that in the RFCD both electrodes are cathodes, and the plasma of the central discharge region appears as the effective anode, being under the highest potential in the discharge.

A very significant difference is that in the discharge gap of the DCD there can not be a potential difference greater than the externally applied voltage, and in the RFCD the instantaneous potential difference in the NESCL can be greater than the amplitude of the applied RF voltage ($V(t) > V_\sim$). In the highly asymmetrical

RFCD, the instantaneous potential difference in the NESCL can approach the doubled amplitude of the RF voltage ($V(t) \leq 2V_\sim$).

The principal difference in the properties of the NESCL of the DCD and RFCD is that in the first case only the conduction current flows in the layer, and in the second case there the conduction and bias currents.

If only the electric field from the external source is present in the NESCL of the DCD, then in the NESCL of the γ-type RFCD there is a superposition of two types of the field – the external RF field and the quasi-stationary intrinsic polarization field of the space charges. As a result, in the latter case, unipolar voltage pulses arise in the NESCL.

The parameters of the NESCL of the DCD are constant in time, and in the RFCD they change with the field frequency and a short-term collapse of the NESCL is possible.

In the DCD, the energy spectrum of ions arriving at the cathode is determined by the external parameters of the discharge (pressure and type of gas, applied voltage) and remains constant for a fixed set of parameters. At the RFCD with the same set of parameters mentioned, with a change in the frequency of the RF field, the form of the energy spectrum of the ions arriving at the electrodes varies greatly.

The parameters of the electron emission processes at the cathode of the DCD, caused by ion bombardment, as well as the electron beam formed in this region in the cathode drop region, are constant in values and in time under the fixed experimental conditions. In the RFCD, because of what was said earlier, with other fixed experiment parameters with a change in the frequency of the RF field the parameters of the emission processes and the near-electrode electron beam as well as their time dependence change significantly.

It should be noted that, due to the edge effect of the RF field in the vicinity of the electrodes, the emerging radial field component enters the space charge layers near the dielectric walls of the discharge tube. There, due to the phenomenon of RF detection (described below), a potential difference greater than the 'floating' potential ($U_s \gg U_{sfl}$) should appear in these layers. As a result, in addition to near-electrode electron beams, in the short discharge gaps beams can enter into the discharge from the entire dielectric cavity surrounding the discharge.

A very important difference in the properties of the DCD and RFCD is that, unlike the DCD, in the RFCD the 'NESCL–plasma' interface undergoes an oscillatory displacement in space. This circumstance leads to two very significant physical consequences: 1) stochastic heating of the plasma electrons and 2) the possible discrete mechanism of transport of plasma electrons to the electrodes, considered below.

The difference in the behaviour of the NESCL in the RFCD and the DCD leads to the previously mentioned curious effect. In the DCD, positive ions leave the discharge gap to the cathode, and electrons to the anode. In a symmetrical RFCD, both electrons and ions leave. In a sharply asymmetrical RFCD with an earthed large electrode, electrons leave only through the active electrode of a small area, since a stationary potential barrier for electrons arises in front of the grounded electrode.

In contrast to the DCD, where the NESCL occurs in front of a highly conductive metal electrode, in the RFCD the near-electrode layers can appear in front of the surface of any materials (metals, dielectrics, semiconductors). Here, in addition to the obvious difference in the emissivity of different materials, it should be noted that if the metal electrodes are equipotential, then in the case of materials of a different physical nature, at least with respect to the distribution of the electrostatic potential, this is completely different.

2.2. The functional role of near-electrode and wall layers of space charge in the RFCD

Let us consider what functions are performed by the near-electrode and wall layers of the space charge (SCWL) in the RFCD.

First of all, it should be noted that the layers of space charge (SCL) before any surface in contact with the plasma, ensure the stationarity of physical conditions at the 'surface–plasma' interface. In this case, a stationary jump in the potential U_s, which slows down the electrons and accelerates the positive ions, is established self-consistently in the SCL and, as a result, equal charges of the opposite sign arrive at a unit area of the surface. This applies equally to both the NESCL and the SCWL.

The oscillogram of the charge fluxes to the electrode, shown in Fig. 2.1, shows the area of the ion and electronic sections of the oscillogram equal for the period of the RF field, which means that the charges of different sign are equal.

Due to the non-linear CVC of a NESCL similar to that of a diode, the phenomenon of RF voltage rectification occurs in the layer,

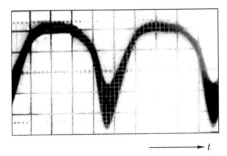

Fig. 2.1. Oscillogram of charge fluxes to the electrode in RFCD, He, $p = 0.1$ Torr, $f = 2$ MHz, $V_\sim = 700$ V.

leading to a quasi-stationary jump in the electric potential U_s in the NESCL of the order of the amplitude $V_\sim$ of the applied RF voltage.

In the RFCD of the γ-type, the NESCL provides the formation of a high-energy electron beam, which is the main ionizing factor and enriches the electron energy spectrum of the plasma with high-energy electrons.

The oscillating outer boundary of the NESCL creates an additional stochastic heating of plasma electrons.

In wall SCLs, due to the edge effect of the RF field, wall electron beams directed to the axial region of the discharge are formed and return some of the energy of the RF field leaving the discharge gap in the radial direction.

The radio-technical impedances of the NESCL in RFCD are the shoulders of the RF voltage divider in the discharge gap, performing the role of ballast resistances.

The NESCL determines the time course of physical processes in the RFCD due to oscillation of their width up to complete collapse.

Depending on the experimental conditions, after the RF gas breakdown, the NESCL is self-consistently set with parameters that determine the type of RFCD that emerges (α- or γ-discharge).

According to the literature data [188], in the RFCD the NESCL together with the near-electrode plasma can form a series resonance oscillatory circuit with a resonance frequency below the plasma frequency ω_{e0}, but substantially higher than the frequency of the RF field supporting the discharge.

2.3. The mechanism of the conductivity of discharge current in the NESCL

The mechanism of the conductivity of the discharge current in the

NESCL significantly affects the physical mechanism of RFCD as a whole.

There are certain traditional ideas on this issue, which are still often used without sufficient grounds for creating physical models of RFCD. A brief set of assumptions is as follows:

1) the width of the NESCL d_s during the period of the RF field varies from the maximum value of $d_{s\ max}$ to the complete disappearance of the layer $d_{s\ max} \geq d_s \geq 0$;

2) the NESCL does not contain electrons;

3) the active conductivity of the NESCL is determined only by the motion of the ions and is much less than the reactive conductivity in magnitude;

4) the 'NESCL–plasma' interface is quite sharp, with a spatial length not exceeding the Debye radius r_{De};

5) in each period of the RF field the plasma enters a short-term contact with electrodes;

6) the dependence of the conductivity of the NESCL on the frequency f and the amplitude of the RF voltage $V_\sim$ is not taken into account;

7) on the equivalent circuit of the RFCD the near-electrode layer is represented in the form of a capacitance.

As a result of the conducted studies, it seems necessary to change the nature of the assumptions as follows:

1) it is necessary to separately consider the RFCD in α- and γ-modes with different equivalent electrical circuits, where the NESCL of the α-discharge is represented by one capacitance C_s, and the NESCL of the γ-discharge by capacitance C_s and the active resistance R_s in parallel;

2) for most RFCD regimes, the plasma does not touch the electrodes at any time in the RF field, as can be seen, for example, from the time dependence of the spatial structure of the near-electrode region (Fig. 1.55);

3) electrons moving in different directions are present in the NESCL during practically the entire period of the RF field T: these are the electrons emitted by the electrode, the electrons produced during the ionization in the layer and coming from the plasma in the 'anodic' phase of the electrode at the minimum height of the potential barrier in the NESCL;

4) taking into account what was said in point 3), it can be assumed that the actual active conductivity of the NESCL significantly exceeds that created solely by the motion of the ions;

5) due to the relatively low density of electrons in the layer ($n_{es} \ll n_{is}$), the NESCL exists for most of the period of the field T;

6) in accordance with clause 5), a near-electrode electron beam significantly modulated in density n_{eb} and energy ε_{eb} also exists for a large part of the period T;

7) after the $\alpha \rightarrow \gamma$-jump on the I–V characteristic, the active conductivity of the NESCL increases by one or two orders;

8) the ratio of the active and reactive components of the discharge current greatly depends on the amplitude $V_\sim$ and the frequency f of the RF voltage;

9) the experimental results considered below give grounds for assuming that at elevated RF voltages the 'NESCL–plasma' interface has a spatially extended ($\Delta x \gg r_{De}$) and structured character.

We note some data on the nature of the conductivity of the discharge current in the RFCD. Thus, for example, using the γ-branch of the CVC for the RFCD in He (Fig. 1.19 *b*) and measuring the phase shift between the RF current and the voltage at the first and last points of this branch of the normal regime, it can be seen that during this γ-branch the ratio of the active current to reactive current I_a/I_p increases almost three times.

At low frequencies of the RF field f, the contribution of the active conductivity is increased, since the bias current is small ($I_{\sim p} = CfV_\sim$), and the amplitude of charge oscillations is large ($A_{e,i} \sim 1/f$).

When the frequencies of the field f are high, the bias current is high in the NESCL and reactive conductivity predominates. At the same time, since the charge density also increases $n_e \sim f^2$, the conduction current (active conductivity) also increases.

The features of the effect of ions on the characteristics of RFCD are seen, for example, from the experimental results of Ref. [194], in which attention was given to the frequency dependences of the minimum discharge-maintaining voltage (Fig. 2.2) and the discharge-maintaining voltage $V_\sim$ and the quasi-stationary voltage U_s in the NESCL with a constant density of the discharge RF power (Fig. 2.3).

The explanation of the dependence $V_\sim(f)$ in Fig. 2.2 is as follows: the voltage $V_\sim$ decreases with increasing frequency f, since at the same time the losses of charged particles due to their vibrational motion decrease as well as due to the fact that at low frequencies the field in the NESCL transfers energy not only to the electrons but also to massive ions, and at high frequencies only to the electrons supporting ionization in the volume.

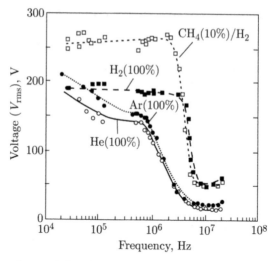

Fig. 2.2. Dependence of the minimum RF voltage supporting the discharge, on the frequency of the field. RFCD in gases – Ar, He, CH_4 (10%) / H_2, H_2, $p = 1$ Torr [194].

It was established that the ion plasma frequency f_{0i}, characterizing, as is well known, the degree of inertia of the ions, is the limiting frequency f of the abrupt transition to lower values of the voltage $V_\sim$ at low gas pressures (collisionless NESCL).

At elevated pressures (collisional NESCL), the experiment shows that the frequency $f_{ITF} = 1/\tau_{ITF}$ is the limiting frequency for a significant decrease in RF voltage, where τ_{ITF} is the ion transit time of the NESCL in the mobility mode becomes of the order of the duration of the RF field period (ITF – ion transit frequency).

The dependence $V_\sim(f)$ for a fixed power input to the discharge reveals an even more pronounced anomaly in the vicinity of the frequency f_{ITF} (Fig. 2.3). As can be seen from the graph, the amplitude of the RF field $V_\sim$ decreases with increasing frequency, which is understandable as a general trend. However, in the vicinity of the frequency $f \sim f_{i0}$ (or at an elevated pressure $f \sim f_{ITF}$), a substantially non-monotonic behaviour of the dependence $V_\sim^{P_a=\text{const}}(f)$ is observed. In this section of this dependence, the amplitude of ion oscillations becomes sufficiently small $A_i \sim 1/f$ and the current of active conductivity (ionic) decreases sharply. In this case, the contribution of the bias current to the full RF current sharply increases, and it is found that the phase shift between the RF current and the voltage $\varphi \approx \pi/2$, i.e., the conductivity is close to reactive. To maintain a constant value of the active power input $P_a = \text{const}$, it is necessary to increase the RF voltage. However,

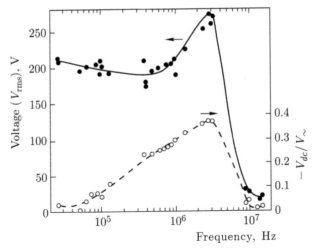

Fig. 2.3. The dependence of the RF voltage and the quasi-stationary potential jump in the NESCL at a constant RF power density $P_{\sim} = 3$ mW $\cdot$ cm^{-2} on the frequency of the RF field. RFCD, Ar, $p = 1$ Torr [194].

further with a monotonic increase in the field frequency the effect of increasing the charge density $n_e \sim f^2$ becomes dominant, and again the active conductivity makes a decisive contribution to the total current. Here it should be noted, that in this experiment the absolute value of the characteristic frequency $f \sim f_{i0}$ depends on the input power $P_a = $ const.

The obtained dependence $U_s(f)$ (Fig. 2.3) is not discussed in Ref. [194]; however, it is explained below on the basis of an analysis of the ion motion in the NESCL in [195].

2.4. The phenomenon of RF detection in the NESCL

In the sixties of the last century interest in the determination of the potential difference of the quasi-stationary field U_s in the NESCL between the electrode under the RF potential and the adjacent plasma was manifested. In [28], the mechanism of occurrence of the U_s voltage (the phenomenon of 'RF detection') in the vicinity of the electrode under the RF potential, in contact with the plasma of the positive column of the DCD, was considered in detail and U_s measurements were taken as a function of the amplitude of the applied RF voltage.

Briefly, the mechanism of the RF detection phenomenon, represented graphically in Fig. 2.4, is as follows.

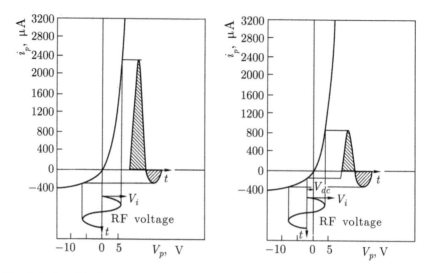

Fig. 2.4. Graphical representation of the mechanism of the phenomenon of 'RF detection' [28].

An RF voltage with an amplitude $V_\sim$ is applied to the positive ions of the NESCL between the electrode and the plasma, which has a non-linear CVC, analogous to the corresponding characteristic of the diode. As can be seen from Fig. 2.4, at initial instants of time, different electron and ion current pulses enter the electrode. As a result, due to the excess of the negative charge entering the electrode in the first periods of the RF field, a negative surface charge arises that retards the electrons and accelerates the ions. After a while, the pulses of the electron and ion currents to the electrode are compared in magnitude, and the stationary potential difference U_s is established in the layer called in the literature the 'self-bias' of the surface potential with respect to the potential of the plasma.

Using the earlier approach [28, 196] and representing the electrodes of a symmetric RFCD as a double probe, the authors of Ref. [197] obtained an expression for U_s in a collisionless NESCL.

A procedure useful from the methodological point of view for calculating the jump in the potential U_s [197] will be described.

When considering the currents on the electrodes, it was taken into account that the potentials of both electrodes are in antiphase, and charges move to the electrodes alternately, then in accelerating, then in the retarding field. The instantaneous value of the discharge current $i(t)$ is equal to the currents $i_A(t)$ and $i_B(t)$ on the electrodes with areas A and B, respectively.

We write the expressions for the instantaneous currents on the electrodes using the RF voltage normalized to the electron temperature (in eV):

$$i(t) = A\left[j_i - j_e \exp\left(\frac{V_A(t)}{V_e}\right)\right] = Bj_e \exp\left(\frac{V_B(t)}{V_e}\right), \tag{2.1}$$

$$V_e = \frac{kT_e}{e}, \quad j_i = en_i\sqrt{\frac{3kT_i}{M}}, \quad j_e = en_e\sqrt{\frac{3kT_e}{m}}, \tag{2.2}$$

$$V_A(t) - V_B(t) = V_\sim \sin \omega t + \varphi_0,$$
$$V_B(t) = V_A(t) - V_\sim \sin \omega t - \varphi_0. \tag{2.3}$$

From (2.1) taking (2.2) and (2.3) into account, one obtains

$$\frac{j_i}{j_e} = \exp\left(\frac{V_A(t)}{V_e}\right)\left[1 + \frac{B}{A}\exp\left(-\frac{V_\sim \sin \omega t + \varphi_0}{V_e}\right)\right]. \tag{2.4}$$

Further, assuming that the plasma is isothermal ($T_e = T_i$) for simplicity, which does not lead to a significant error, since $\dfrac{j_i}{j_e} \sim \sqrt{\dfrac{T_i}{T_e}}$, it is easy to obtain from (2.4) the expression for the instantaneous jump of the potential in front of the electrode A:

$$V_A(t) = -V_e \ln\sqrt{\frac{M}{m}} - V_e \ln\left[1 + \frac{B}{A}\exp\left(-\frac{V_\sim \sin \omega t + \varphi_0}{V_e}\right)\right]. \tag{2.5}$$

To obtain the expression for the quasi-stationary jump in the potential $\bar{V}_A(t)$ in front of the electrode A, we average (2.5) over one period of the field 2π:

$$\bar{V}_A(t) = -V_e \ln\sqrt{\frac{M}{m}} - \frac{V_e}{2\pi}\int_0^{2\pi} \ln\left[1 + \frac{B}{A}\exp\left(-\frac{V_\sim \sin \omega t + \varphi_0}{V_e}\right)\right]d(\omega t). \tag{2.6}$$

To further simplify the expressions, we will consider the RFCD to be symmetric: $A = B$, then $\varphi_0 = 0$, and we get:

$$\bar{V}_A(t) = -V_e \ln\sqrt{\frac{M}{m}} - \frac{V_e}{2\pi}\int_0^{2\pi} \ln\left[1+\exp\left(-\frac{V_\sim \sin\omega t}{V_e}\right)\right]d(\omega t),$$

whence after integration the final expression for the quantity $\bar{V}_A(t)$ takes the form:

$$\bar{V}_A(t) = -V_e \ln\sqrt{\frac{M}{m}} - \frac{V_\sim}{\pi} + \frac{V_e}{2\pi}. \qquad (2.7)$$

Hence we obtain for $V_\sim \gg V_e$:

$$\bar{V}_A(t) = U_s \approx -\frac{V_\sim}{\pi}. \qquad (2.7a)$$

In the limiting case of the absence of RF voltage ($V_\sim = 0$) formula (2.7) takes the form:

$$U_s \approx -V_e \ln\sqrt{\frac{M}{m}} = \varphi_{fl},$$

which coincides with the well-known expression for the 'floating' potential φ_{fl}.

2.5. Dependence of the quasi-stationary jump in the potential U_s in the NESCL on the parameters of the RFCD

Let us consider the dependence of the magnitude of the quasi-stationary jump in the U_s potential, as an important characteristic of RFCD, on its main parameters.

2.5.1. The patterns of U_s versus the amplitude of the RF voltage $V_\sim$

A characteristic form of the spatial distribution of the quasi-stationary electric potential $\varphi_0(x)$ observed in a symmetric RFCD with a short discharge gap is shown in Fig. 2.5. Jumps of the potential U_s in the near-electrode region can be seen.

The results of a study of the dependence of the value of U_s on the applied RF voltage for various conditions are shown in Figs. 2.6–2.10.

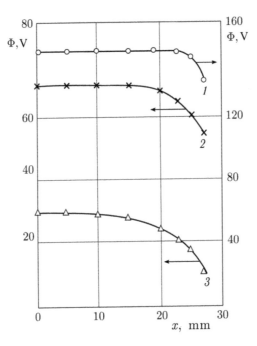

Fig. 2.5. Distribution of the quasi-stationary potential of the electric field in the discharge gap of the RFCD. He, p = 0.5 Torr, f = 5 MHz . $1 - V_\sim$ = 300 V, $2 - 120$ V, $3 - 40$ V.

As can be seen from the obtained results, the monotonic increase in U_s observed with many authors [23, 27] with increasing $V_\sim$ is confirmed, as well as the commensurability of these quantities.

We note that at relatively low RF voltages and gas pressures ($V_\sim \leq 500$ V, $p < 1$ Torr), the dependence $U_s(V_\sim)$ is practically linear. In this case, the angle of inclination of this straight line essentially depends on other discharge parameters.

In the case of increased voltages and pressures ($V_\sim > 500$ V, $p \geq 1$ Torr), there was a tendency to slow the growth of the $U_s(V_\sim)$ curve (Figs. 2.7, 2.8). The latter can be explained by two causes. First, at increased voltages $V_\sim$, the width of NESCL decreases because of increased capacity of NESCL and decreased impedance. Then, the RF voltage drop at NESCL is smaller, and the value of RF voltage detected in NESCL decreases. Second, by increasing the emission of electrons from the electrode under the action of increasing bombardment by ions. Such a discharge of negative charge from the surface of the electrode, in accordance with the theory of the phenomenon of 'RF detection' [28], should also lead to a decrease in U_s.

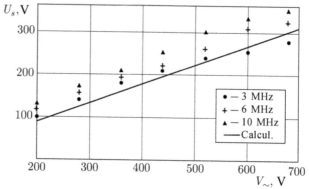

Fig. 2.6. A comparison of the measured magnitude of the quasi-stationary jump in the potential U_s in the NESCL with the calculation in [197], RFCD in He at $p = 0.5$ Torr.

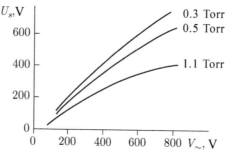

Fig. 2.7. Dependence of the value of the jump in the potential U_s on the amplitude of the RF voltage $V_{\sim}$ in the RFCD. Ne, $f = 6$ MHz, the electrodes are Ti.

A comparison of the experimental data with the theory is shown in Fig. 2.6, which indicates that for a field frequency of 3 MHz there is good agreement, and at a frequency of 10 MHz, the discrepancy already reaches 25%. This is due to the approximate nature of the theory, which does not take into account the dependence of U_s on the frequency of the field and the collisions of particles in the layer.

Later in [198] an expression was obtained for U_s, taking into account the dependence on the field frequencies and collisions of electrons with atoms in the layer:

$$U_s = V_e \ln \sqrt{\frac{M_i}{2\pi m}} + \frac{1}{\pi}\left[\frac{\omega}{\nu}V_p + \left(V_{\sim}^2 - V_p^2\right)^{1/2}\right], \qquad (2.8)$$

where V_p is the voltage drop across the plasma, ω is the circular frequency of the RF field, and ν is the effective collision frequency of electrons with atoms.

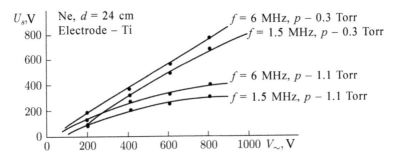

Fig. 2.8. Dependence of the value of U_s on the voltage $V_{\sim}$.

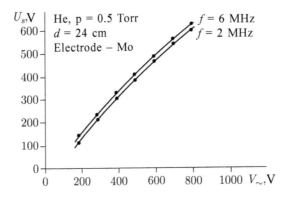

Fig. 2.9. Dependence of U_s on the voltage $V_{\sim}$ for two frequencies of the RF field.

Let us compare the experimental value of U_s calculated for the case presented in Fig. 2.9. Here, at $V_{\sim} = 400$ V, the experimental value is $U_s = 300$ V, and calculated value using the reduced expression $U_s \approx 182$ V. It is clear from this that the theoretical values of U_s are about 40% smaller than the experimental values.

Thus, the expressions available in the literature so far allow us to estimate with the specified accuracy U_s, but not to calculate their exact values.

It should be noted that the measured values of U_s in the RFCD in the investigated gases with the use of various electrode materials reached and exceeded the table values of the normal cathode drop voltages in a glow discharge dc with similar external parameters [151].

The foregoing suggests the possibility of the existence in the discharge gap of the RFCD of an additional discharge of direct

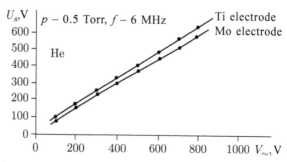

Fig. 2.10. The dependence of U_s on the voltage $V_\sim$ for different electrode materials.

current, the characteristics of which have been observed and illuminated below, and also noted in the literature [190, 199].

2.5.2. Dependence of U_s on gas pressure

The results of an experimental study of the dependence of the value of U_s on the gas pressure are shown in Figs. 2.7, 2.8 and 2.11. Apparently, in accordance with the data of [23] and the mechanism of the phenomenon of 'RF detection' [28], the jump in the potential U_s monotonically decreases with increasing p. However, in contrast to [23], in our work the measured values of U_s remained large enough up to the maximum used values $p = 1.1$ Torr.

The measurements carried out in Ref. [200] showed that in the range of average pressures ($1 \le P \le 100$ Torr) the jumps in the potential U_s are tens and hundreds of volts. This is due to the fact that the thickness of the near-electrode layer d_s decreases with increasing pressure, and despite the considerable decrease in the mean free path of electrons and ions, the conditions for realizing the mechanism of the phenomenon of 'RF detection' are preserved.

2.5.3. Dependence of U_s on the material of the electrode and the kind of gas

As far as is known, the dependence of the jumps in the potential U_s in RFCD on the material of the electrodes and the kind of gas was firstly studied in [195, 201]. Thus, Fig. 2.10 shows the $U_s(V_\sim)$ curve obtained during measurements in a helium discharge with electrodes made of titanium and molybdenum. A similar experimental curve is shown for the RFCD in Ne also for the Ti and Mo electrodes in Fig. 2.12.

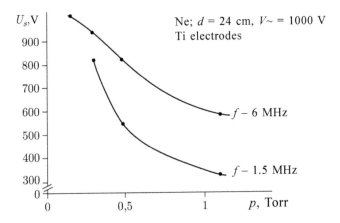

Fig. 2.11. Dependence of the jump in the potential U_s on the gas pressure p for different field frequencies.

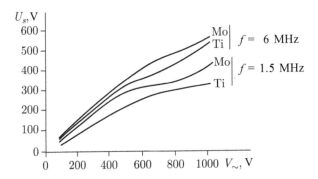

Fig. 2.12. Dependence of the value of the jump in the potential U_s on the material of the electrodes of the RFCD in Ne, $p = 1.1$ Torr.

The $U_s(V_\sim)$ dependence for the discharge with the molybdenum electrodes in helium and neon is illustrated in Fig. 2.13.

As can be seen from the obtained results, the values of U_s depend essentially on the material of the electrode and the kind of gas.

In the observed dependence, two mechanisms should be distinguished: (1) a change in the intensity of the process of electron emission from the surface of the electrode; 2) the change in the parameter m/M_i, which depends on the ratio of the degrees of inertness of electrons and ions, which determines the magnitude of the jump in the potential U_s according to the physical model of the phenomenon of 'RF detection' [28].

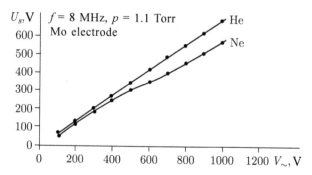

Fig. 2.13. Dependence of the value of the jump in the potential U_s on the nature of the gas.

It is well known that the parameters of the electron emission process are determined by the combination of the 'electrode material–type of gas' pair [151].

The work function φ of the emitter material determines the intensity of all emission mechanisms. According to the data of [202], for Mo and Ti we have: $\varphi_{Mo} = 4.09$ eV, $\varphi_{Ti} = 3.72$ eV.

The results in Fig. 2.12 show that under other fixed conditions in the discharge, the value of U_s for the molybdenum electrodes is higher than for titanium. If we take into account that stronger emission corresponds to a lower value of U_s, then the experimental data correspond to the ratio of the work functions φ_{Mo} and φ_{Ti}.

The dependence of U_s on the type of gas, in connection with emission processes, is due to the fact that the generalized electron emission coefficient γ from the electrode surface is determined by the action of ions, photons, excited atoms and fast atoms arising during the process of charge exchange on the electrode. The above factors of the effect of the plasma on the electrode will naturally be different for different gases.

The influence of the gas type on the characteristics of a DC glow discharge, associated with emission processes, including the normal cathode drop V_{cd}, was investigated in Ref. [202]. In particular, for the molybdenum cathode it has been established that V_{cd} for fixed discharge parameters decreases monotonically for the following sequence of gases: He, Ne, Ar.

As can be seen from Fig. 2.13, the U_s values for the He–Mo combination are higher than for the Ne–Mo pair, which qualitatively corresponds to the above-mentioned course of the dependence of V_{cd} on the type of gas.

We also note the mechanism of the dependence of U_s on the type of a gas of the non-emission character.

The phenomenon of 'RF detection' (RFD), which creates, mainly, jumps in the potential U_s, arises in the case of a large difference in the mobility of charges of the opposite sign. The latter means that the parameter $m/M_i \ll 1$ is small. Thus, a decrease in the mass of the ion M_i causes a tendency to decrease the value of U_s.

In Fig. 2.13, however, the U_s values for the light gas (He) exceed the corresponding values for the heavier gas (Ne). Consequently, in this case, other mechanisms have a predominant significance in establishing the value of U_s.

At the same time, we note that for relatively low values of RF voltage $V_\sim$, when U_s is of the same order as the 'floating' potential of

$$V_{fl} = V_e \ln \sqrt{\frac{M_i}{m}},$$ the jumps of the potential for both gases are close

in magnitude. They are brought together by opposite tendencies. In the range of values of $U_s \gg V_{fl}$, the measured values of U_s for He and Ne already significantly diverge (Fig. 2.13).

The dependence of the near-electrode quasi-stationary fields of RFCD on the choice of the 'electrode material–kind of gas' pair was also confirmed in work [190].

2.5.4. The frequency dependence $U_s(f)$

The dependence of the jump in the potential U_s on the frequency of the RF field in the range $1 \le f \le 25$ MHz with other discharge parameters unchanged was studied in experiments.

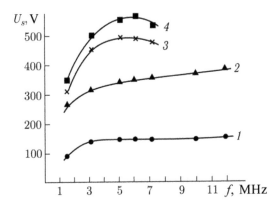

Fig. 2.14. Dependence of the jump in the potential U_s on the frequency of the RF field. Ne, $p = 1$ Torr, Ti electrode, $V_\sim = 200$ (*1*), 600 (*2*), 900 (*3*), 1000 (*4*) V.

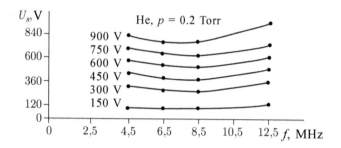

Fig. 2.15. Dependence of the jump in the potential U_s on the frequency and voltage of the RF field, He, $p = 0.2$ Torr.

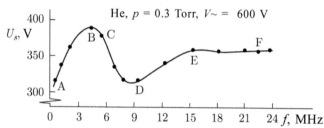

Fig. 2.16. The dependence of U_s on the field frequency in the extended frequency range.

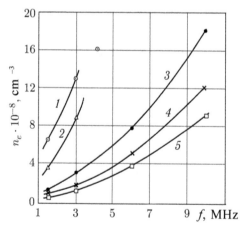

Fig. 2.17. Dependence of the electron density n_e of the RFCD plasma on the frequency and the RF field voltage. He, $p = 0.5$ Torr. $1 - V_\sim = 480$ V, $2 - 400$ V, $3 - 260$ V, $4 - 200$ V, $5 - 160$ V.

The experimental results obtained are shown in Figs. 2.6, 2.14–2.16.

Let us analyze the experimental data, noting that in the literature there are no similar experimental studies of RFCD.

The theoretical expression for U_s in [198], which includes the frequency of the field, is not suitable for the conditions studied because of the condition $V_{\sim} \gg V_p$ (V_p is the drop in the RF voltage on the plasma)is fulfilled. In this case, the member containing the field frequency is small $\left(\dfrac{\omega}{\nu}\right)V_p \ll V_{\sim}$ so that the frequency dependence practically disappears.

Analysis of the experimental data is complicated by the fact that, for other fixed discharge parameters, there is a strong dependence of the concentration of plasma electrons on the field frequency. A typical example of such a dependence in the investigated conditions is shown in Fig. 2.17.

Analysis of the experimental results (Fig. 2.14), obtained in the frequency range (2–12.5 MHz), shows that a monotonous increase in the amplitude of the RF voltage leads to three characteristic cases of the frequency dependence of $U_s(f)$.

The first case of the dependence $U_s(f)$ corresponds to low RF voltages ($V_{\sim} \leq 400$ V), when U_s increases from the minimum value at $f \sim 1$ MHz to some constant value.

With increasing voltage ($400 \leq V_{\sim} \leq 800$ V) the monotonous growth of U_s was observed in the region of high frequencies.

Finally, the third case is realized for $V_{\sim} \geq 800$ V, when the shape of the curve $U_s(f)$ is non-monotonic.

As already noted, the essence of the phenomenon of RF detection (RFD) is related to the different character of the motion of electrons and ions in rapidly varying fields and to various degrees of inertia of these particles.

It is necessary to clarify the physical meaning of the manifestation of the inertness of charged particles which in this case is understood as the ability of a charged particle to reach the electrode surface during the half-period of the RF field $T/2$, favourable for a given particle in the sense of advancing to the electrode.

According to the RFD mechanism model [28] the value of U_s is determined to a large extent by the density of the negative charge on the surface of the electrode. The positive ions, reaching the electrode, reduce the surface charge and, consequently, the value of U_s.

Let us analyze the low-frequency region when U_s increases with increasing f.

One of the reasons for the $U_s(f)$ dependence may be a change in the nature of the penetration of the RF field into the plasma and, correspondingly, the change in the voltage drop across the NESCL.

The impedance of a capacitor filled with plasma in the presence of space charge layers was investigated in [203–205]. The approach used in these works can be applied to the analysis of processes in RFCD.

According to the results of the investigation of the 'plasma' capacitor, in the case of low and medium frequencies ($f^2 \ll f_{0e}^2$), the nature of the penetration of the external longitudinal RF field into the plasma depends on the ratio of the field frequency f and the ion Langmuir frequency f_{0i}.

With $f \leq f_{0i}$, the impedance of the plasma filling the capacitor is commensurate with the NESCL impedance, and some of the applied RF voltage falls in the region occupied by the plasma.

When does the condition $f^2 \gg f_{0i}^2$, the plasma impedance is small compared to the impedance of the layers, and the applied RF voltage almost completely falls in the layers. In this case the value of U_s should reach the maximum value.

According to the experimental data shown in Fig. 2.16, the increase in U_s with increasing f for a fixed $V_\sim$ can be explained by the transition of conditions in the discharge from the case $f \sim f_{0i}$ to the case $f^2 \gg f_{0i}^2$.

The obtained curves of the frequency dependence $U_s(f)$ show that there exists a characteristic frequency f_s with which saturation or a significant slowdown in the growth of U_s begins with low frequencies, with a further increase in frequency (Fig. 2.14).

The following physical model is proposed, which explains the appearance of the boundary frequency f_s, and an expression for its calculation.

Recall that, reaching the surface of the electrode, the positive ion reduces the negative surface charge and the value of U_s. With increasing frequency of the field f, the inertia of the ions increases with respect to the advance to the electrode for each half-period of the field accelerating ion.

The mean free path of the ion λ_i in the conditions under consideration is determined by the charge exchange process [157]:

$$\lambda_i = \frac{1}{P_t p},$$

where P_t [cm^{-1}/Torr] is the probability of charge exchange, p [Torr] is the gas pressure.

We consider a region adjacent to the electrode with a length of one free path λ_i, in which the ions move without collisions. Obviously, the ion travels through this region in the uniformly accelerated motion during the time t':

$$\lambda_i = \frac{1}{P_i p} = \frac{eE}{2M_i}(t')^2,$$

where E is the intensity of the quasi-stationary field in the neighbourhood of the electrode. From here:

$$t' = \left(\frac{2M_i}{eEP_i p}\right)^{1/2}.$$

Thus, the maximum frequency f_s at which the ion still has time to pass the distance λ_i in a half-period of the field $T_s/2$, is determined by the expression:

$$t' = \frac{T_s}{2} = \frac{1}{2f_s} = \left(\frac{2M_i}{eEP_i p}\right)^{1/2}$$

or

$$f_s = \left(\frac{eEP_i p}{8M_i}\right)^{1/2}.$$

Let us estimate the frequency f_s for the experimental conditions (Fig. 2.14, $V_\sim = 200$ V):

$$U_s = 140 \text{ V}, \ d_s \sim 1 \text{ cm}, \ E = 140 \text{ V/cm}, \ P_t \approx 70\,[\text{cm}^{-1}\cdot\text{Torr}^{-1}]\,[4],$$

$$\text{Ne}, \ p = 1.1 \text{ Torr}.$$

$$f_s = \left(\frac{4.8 \cdot 10^{-10} \cdot 140 \cdot 70 \cdot 1.1}{8 \cdot 1.68 \cdot 10^{-24} \cdot 20 \cdot 300}\right)^{1/2} \approx 7 \text{ MHz}.$$

The experimental value of $f_s^{\text{exp}} \approx 5.5$ MHz. Taking into account the approximate nature of the values of E and P_t used in the calculation, we can assume that the correspondence between the values of f_s^{calc} and f_s^{exp} is satisfactory.

In the low-frequency region under consideration ($f \leq 5$ MHz), the mechanism that decreases U_s increases with increasing f, since the concentration of charged particles in the plasma and, consequently, the density of the ions bombarding the electrode increase. The latter increases the intensity of the emission of electrons from the electrode, which decreases U_s.

Since the increase of U_s has been experimentally recorded with increasing f, it is obvious that the contribution of the effect of an increase in the inertia of ions predominates over the contribution of emission processes in the formation of a jump in the potential U_s.

As can be seen from Fig. 2.14, in the range of RF voltage $400 \leq V_\sim \leq 800$ V increasing frequency also increases U_s monotonically. According to estimates for these conditions, we obtain: $f > 12$ MHz. Thus, the saturation of the dependence $U_s(f)$ occurs at frequencies exceeding the maximum frequency used in the given measurements.

In the case of $V_\sim > 800$ V the behaviour of the curve $U_s(f)$ is non-mootonic, which was also noted in a number of other experiments (Figs. 2.15 and 2.16).

The frequency dependence $U_s(f)$ in RFCD in helium for a somewhat wider range is shown in Fig. 2.16. From this it is seen that after the section of the non-monotonic dependence on f, the value U_s becomes practically constant.

Let us analyze the obtained dependence $U_s(f)$, selecting the characteristic points (A–E) on it and explaining their significance (Fig. 2.16). Point A is the first experimental point; B is the point of the first extremum of the curve; C – indicates the position of frequency equality $f = f_{0i}$; D is the point of the second extremum; E indicates the position of the characteristic frequency f_s.

We note the processes that determine the value of U_s under the conditions under consideration.

According to what was said above, in the frequency range AC we have $f_{0i} \geq f$, as a result of which the RF field penetrates into the plasma, and the applied RF voltage is distributed in the discharge gap between the NESCL and the plasma.

Accordingly, the condition $f > f_{0i}$ is satisfied in the section CF, and the field is increasingly localized in the NESCL, increasing U_s.

Electronic emission, proportional to the flux of ions to the electrode, reduces U_s. The ion flux essentially depends on the field frequency, with the growth of which the ion density increases and the mode of their motion to the electrode changes. At low frequencies, the motion of ions is determined by their mobility in the electric field, and at higher frequencies ($f > f_{0i}$) – diffusion to the electrode.

On the section AB of the curve $U_s(f)$ (Fig. 2.16), the effect of emission is small, the effects increasing U_s predominate – the near-electrode jump in the potential increases. The section BD differs by the prevalence of the emission process, which reduces U_s. At frequencies $f > f_D$, the motion of ions occurs in a slowed down diffusion mode, and the localization of the RF field in the NESCL increases. Therefore, an increase in the value of U_s is observed in the range DE. After frequency $f_E \approx 15$ MHz, the value of the U_s jump is stabilized. An estimate of the frequency at which the $U_s(f)$ dependence is saturated yields a value $f_s = 14$ MHz, which is very close to the frequency f_E following from the experiment.

Thus, to the frequency $f_E = f_s$ the processes of establishing the total inertia of the ions and the localization of the RF field in the NESCL are practically completed, the value of U_s ceases to depend on the frequency of the field, and a quasi-neutral stationary flux of charged particles arrives at the electrode.

2.6. The near-electrode jump in the potential U_s in RFCD with external electrodes

Despite the urgency of the practical application of RFCD, the electrodes of which are not in contact with the plasma-forming medium (including chemically aggressive ones), this type of discharge has not been studied experimentally enough.

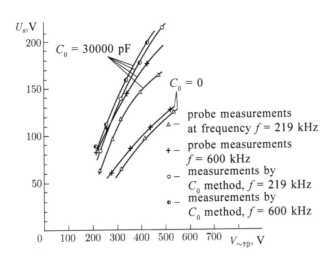

Fig. 2.18. The results of measurements of U_s by the probe and contactless methods in RFCD in He at $p = 0.5$ Torr. V_0 – RFCD mode (capacitance in the discharge circuit $C_0 = 3 \cdot 10^4$ pF); I_0 – RFCD mode (capacitance C_0 in the discharge circuit was absent).

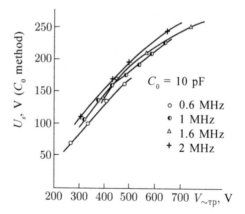

Fig. 2.19. The results of measurements of the value of U_s by the contactless method in RFCD with external electrodes. He, $p = 0.3$ Torr.

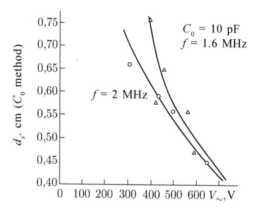

Fig. 2.20. Dependence of the thickness of the NESCL of the RFCD d_s on the amplitude of the RF voltage in the discharge with external electrodes. He, $p = 0.3$ Torr.

In [206], thanks to the developed contactless method, near-electrode jumps of the quasi-stationary potential in RFCD with external electrodes were measured for the first time as a function of the amplitude of the RF voltage and the field frequency.

The characteristic experimental results of measuring the quasi-stationary jump in the potential U_s in the RFCD with external electrodes are shown in Figs. 2.18 and 2.19.

The contactless technique of [206] also makes it possible to measure the capacitance of the near-electrode layer C_s. Hence, in the approximation of a flat capacitor, it is possible to calculate the thickness of the NESCL d_s.

The typical values of d_s measured in this way in a discharge with external electrodes are shown in Fig. 2.20.

The obtained experimental results show that in the RFCD with external electrodes the jumps of the potential U_s are close in magnitude to those in a discharge with internal electrodes.

Similar in these discharges are the lengths of the layers d_s and their dependence on voltage $V_\sim$ and the field frequency.

The obtained values of U_s and d_s can be used to determine the intensity of the quasi-stationary field in the NESCL: $E = U_s/d_s$.

Under the investigated conditions, the intensity of the quasi-stationary field E in the NESCL varied within 30–500 V/cm.

From here one can also estimate the energy ε_i of ions arriving at the electrode:

$$\varepsilon_i = E\lambda_i.$$

According to the obtained experimental data, the ion energy was from 2 to 50 eV.

The experimental information obtained on the parameters of the physical conditions in the NESCL is also of interest for the purposes of plasma technology.

2.7. The nature of the resulting electric field in the NESCL of RFCD

According to the foregoing, much attention was paid in the studies to RF and quasi-stationary components of the electric field in the boundary regions of the discharge.

However, it should be stressed that in Ref. [51], attention is first drawn to the need for a detailed study of the resultant field arising in the NESCL due to the superposition of the external RF field and the total quasi-stationary field of volume and surface charges in studying the physical mechanism of RFCD.

The corresponding diagrams of the time variation of the RF potential of the electrode $\varphi_\sim$, the time-averaged electrode potential Φ_0, the plasma potential V_p, and the resultant voltage V_s are shown in Fig. 2.21.

Of greatest interest is the instantaneous potential difference between the plasma and the electrode, resulting from the superposition of the applied RF voltage $V_\sim$ and the quasi-stationary jump in the potential Φ_0.

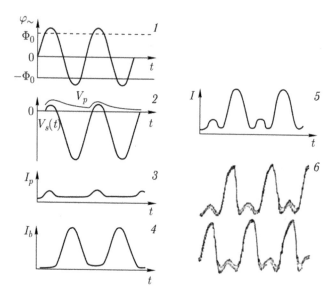

Fig. 2.21. Qualitative construction of the luminescence intensity curve in time in the near-electrode regions of RFCD: *1* – variable ($\varphi_\sim$) and constant (Φ_0) components of the electrode potential, *2* – time variation of the total electrode potential $V(t) = \Phi_0 + \varphi_\sim$, *3* – the time variation of the luminous intensity I_p due to the polarization of the plasma near the electrode, *4* – the time variation of the luminous intensity component I_b due to the pulsating flux of fast electrons, *5* – the resultant curve of the dependence ($I = I_p + I_b$) in the near-electrode region as a function of time, *6* – the oscillogram of the luminescence intensity near both discharge electrodes [57].

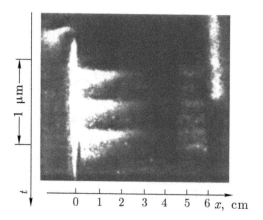

Fig. 2.22. Spatially–temporal change in the glow in the vicinity of the RFCD electrode, He, $p = 0.7$ Torr, $f = 3.3$ MHz, $V_\sim = 2.8$ kV.

As can be seen from Fig. 2.21, the instantaneous potential difference in the NESCL $V_s(t)$ is expressed by unipolar pulses. In this case, the intensity vector of the resultant field $\mathbf{E}(t)$ does not change its direction during the entire period of the RF field, changing in time only in magnitude. The vector $E(t)$ is oriented in such a way that this field in the NESCL accelerates the ions to the electrode, and the electrons emitted by the electrode to the centre of the discharge gap.

Thus, the resulting field in the NESCL is a pulsed constant field, and the RFCD is like a DCGD with a periodic impulse cathode drop.

The following results are the experimental confirmation of the proposed model of physical processes in the NESCL.

The method of superfast photography used in [67] recorded periodic flashes of luminescence, propagating from the electrode to the centre of the discharge (Fig. 2.22).

The above ideas about the nature of physical processes in the near-electrode regions of RFCD have been confirmed in a number of foreign studies [138, 207–209].

Based on the properties of the near-electrode layers, a number of important conclusions can be drawn.

First, it should be assumed that in the region immediately adjacent to the electrode there is no or strongly weakened alternating RF electric field. In this case, the resulting field is a constant field pulse. This result is due to the superposition of the RF field and the quasi-stationary field only within the limits of the NESCL. The residual alternating RF field is restored outside the boundary of the NESCL.

Experimentally, the above-mentioned tendency is confirmed by our studies of the spatial distribution of $I_{2j}(x)$, which drops to the electrode (Figs. 1.41, 1.42).

In addition, direct measurements of $E_{\sim}(x)$ by R. Gottscho [185] also showed a decrease in the field strength $E_{\sim}$ near the surface of the electrode (Fig. 1.37).

It should be noted that the intensity of the effect of attenuation of the RF field in the vicinity of the electrode depends on the ratio of the voltage amplitudes $V_{\sim}$ and U_s, which in turn depends on the gas pressure. With increasing pressure, U_s decreases for a fixed $V_{\sim}$. Therefore, the process of distortion of the RF field by the quasi-stationary field is attenuated.

At low RF voltages, when U_s is small, in the vicinity of the electrode the weakening of the RF field as a result of the superposition of these fields decreases.

The second important conclusion is that when the resulting field in the NESCL acquires the character of constant-field pulses, the motion of the electrons in the neighbourhood of the electrode is not of an oscillatory nature, but becomes pulsed unidirectional from the electrode to the centre of the discharge (Fig. 2.22). In this case, the introduction of the amplitude of the vibrational motion of electrons in the near-electrode region and the conclusions based on this, for example, on the thickness of the NESCL, become incorrect.

Below are various experimental confirmations of the presence of unidirectional motion of emitted electrons in the NESCL.

Third conclusion: since there is no alternating RF field in the RFCD, the notion of 'skinning' of the RF field at the boundary of the gas-discharge plasma and the procedure for determining the depth of penetration of the RF field into the plasma become vague. The situation here is close to the screening of the electrostatic field.

As an experimental confirmation of the absence of the classical 'skinning' of the RF field in the near-electrode region of the RFCD, we give the data of R. Gottscho [97, 185], which show that the spatial distribution of the strength $E_\sim(x)$ does not depend on the frequency of the RF field (Fig. 1.37).

2.8. Features of the properties of NESCL in asymmetric RFCD. The effect of 'self-bias'. 'Battery' effect

As a rule, in academic studies and, in particular, in practice, attention is given to asymmetric RFCD (ARFCD) (Fig. 2.23).

The physical properties of the NESCL of asymmetric RFCD are discussed in detail in [6, 78]. Therefore, let us briefly consider only the main points.

Two ARFCD modes are considered in the theoretical work [78]: the 'self-bias' mode (with a blocking capacitance C_{bl}) and the 'battery' mode (without a blocking capacitance C_{bl}).

In the case of the 'self-bias' mode, in particular, expressions are obtained for jumps of the quasi-stationary potential U_s in the NESCL of the active electrode

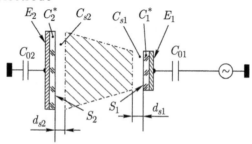

Fig. 2.23. Spatial physical model of asymmetric RFCD.

$$U_{s1} = \frac{2V_{\sim}}{1 + \eta^q + \overline{C}_{s1}/C_{bl}}, \qquad (2.9)$$

where $\overline{C}_{s1}$ is the average capacitance of the layer of the active electrode, $\eta = S_1/S_2$ is the ratio of the areas of the active and grounded electrodes, $q = 1.2–1.5$ is the value from the experiment, which depends on the ion transport mechanism in the NESCL: 1) collisionless NESCL, 2) the collision frequency of ions with neutral particles $v_{in} = $ const; 3) the free path of ions $\lambda_{in} = $ const; U_{s2} – the jump in the potential in the NESCL of a grounded electrode

$$U_{s2} = U_{s1}\eta^q. \qquad (2.10)$$

Since under the experimental conditions: $\eta \ll 1$, $C_{bl} \gg C_{s1}$, we get:

$$U_{s1} \lesssim 2V_{\sim} \quad \text{and} \quad U_{s2} \ll U_{s1}. \qquad (2.11)$$

In addition, as mentioned earlier, in the 'self-bias' mode, the effect of the anharmonicity of the discharge current is strongly manifested.

Consider the essence of the 'battery' effect. For its implementation, the external discharge circuit must be closed in direct current, which is provided by grounding the active electrode through a large inductance. As a result of the analysis of the functioning of the obtained electric circuit ARFCD it was established in [78] that in the process of oscillations the plasma never touches a grounded electrode of a large area, the ionic flux on it proceeds, and the electrons are retained by a stationary potential barrier and do not enter this electrode. In this case, the ion flux also goes to the active electrode with a small area, but it is compensated with excess by the electron flux at the moments of the electrode touching the oscillating plasma.

The 'battery' effect and the effect of 'self-bias' are interrelated: if the external discharge circuit is disconnected by direct current, then a self-bias voltage appears on the terminals of the 'battery' (on the electrodes), playing the role of EMF when the gas-discharge circuit is closed in direct current.

The described effect of 'self-bias' in the ARFCD is very important for the technology of surface treatment, since it allows almost doubling the potential jump in the NESCL U_s before the substrate under processing at a constant RF voltage, introducing only the asymmetry of the discharge.

2.9. Transport of ions of gas-discharge plasma to electrodes

Let us consider the nature of the behaviour of positive ions and their role in the mechanism of RFCD. Let us note in particular the importance of questions on the transport of ions to electrodes, their effect on the surface of the electrode, which causes electronic emission processes, as well as technological processes of modifying the properties of the treated surface.

At first glance, ions play a passive role in the gas discharge and gas-discharge plasma, since the active component of the plasma is electrons that produce ionization and excitation of the neutral gas. In fact, the situation here is ambiguous.

Let us first mention the ion functions common to all the discharges:

1) ions are a plasma-forming component,

2) create layers of space charge at the periphery of the discharge gap (near the electrodes and walls),

3) initiate the emission of electrons from all discharge surfaces,

4) regulate the rate of charge departure from the discharge gap (with the help of SCL or in the process of volumetric recombination in plasma),

5) lead to ion–acoustic oscillations and waves in the plasma,

6) negative ions initiate the excitation of specific instabilities in the plasma,

7) participate in the formation of the current–voltage characteristics of the discharge.

In addition to the ion functions common to various gas discharges, an additional specificity of the ion behaviour and their contribution to the discharge mechanism appears in RFCD:

1) the appearance of the phenomenon of 'RF detection' in the NESCL due to the non-linear I–V characteristic of the ionic layer,

2) the differences in the mechanism of RFCD at low and high field frequencies due to the change in the ion motion regime in the NESCL [194],

3) strong frequency dependence of the energy spectrum of ions on the electrodes [207, 208],

4) the influence of the time and energy distributions of the ions bombarding the electrode on the spatio–temporal behaviour of the plasma electron distribution $f_e(\varepsilon)$ in terms of energy [85].

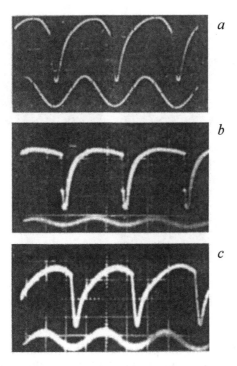

Fig. 2.24. Oscillograms of the fluxes of charged particles to the electrode of a symmetric RFCD. $V_{\sim}$ = 1000 V, f = 1.6 MHz, p = 0.1 Torr. Gases: *a*) helium, *b*) xenon, *c*) oxygen.

The characteristic oscillograms of the fluxes of ions and electrons to the symmetrical RFCD electrode for various working gases and pressures are shown in Figs. 2.24 and 2.25.

On the above oscillograms, the upper part corresponds to the ion current, and the lower part corresponds to the electron current.

As can be seen from Fig. 2.24, the higher stationarity of the ion flux corresponds to the heavier gas (Xe), which is to be expected. The nature of the current pulses of low-inertia electrons is thus adjusted to the transport of ions of each gas.

According to the oscillograms of Fig. 2.25, the process of motion of charged particles to the electrode depends essentially on the gas pressure, both for ions and for electrons. In particular, with an increase in pressure, there is a noticeable change in the ion flux during the period of the RF field.

According to the literature data, up to 80% of the power absorbed in the low-pressure RF discharge can go to the acceleration of ions in the NESCL with high RF voltages.

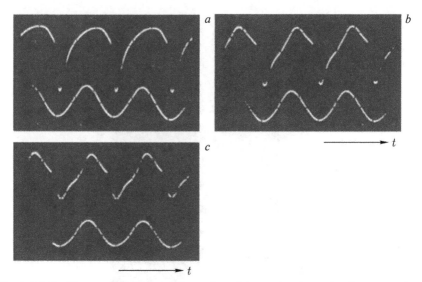

Fig. 2.25. Oscillograms of fluxes of charged particles to the electrode of a symmetrical RFCD for different pressures. Helium. $V_\sim = 1000$ V, $f = 1.6$ MHz. *a*) 0.1 Torr, *b*) 1.0 Torr, *c*) 2.0 Torr.

Due to the fact that the potential of the plasma always exceeds the potential of the electrodes [138, 207, 209], the ion flux modulated in density and energy goes to the RFCD electrodes throughout the entire RF field period. In this case, ions always go to both electrodes, unlike electrons, which in an asymmetric discharge (I_0 – mode) only come to an electrode of smaller area [78, 138].

Considering the frequency range of the examined RFCD ($f_{0i} \leq f \ll f_{0e}$), one should expect the greatest influence of ions on the discharge characteristics at frequencies $f \sim f_{0i}$.

According to what was said above, the vector of the velocity of the ions $\mathbf{v}_i$ in the NESCL is always directed toward the electrodes, that is, the motion of the ions is practically translational rather than oscillatory [51, 138, 208].

In the case of a collisionless NESCL with $f \leq f_{0i}$, the quantity $\mathbf{v}_i(x)$ depends on the instantaneous value of the field strength $\mathbf{E}(x)$.

In the case of the collisional NESCL, the velocity $\mathbf{v}_i$ in the negative half-period of the field contains the drift and diffusion components, while the positive half contains only the diffusion component.

The equation of the ion flux to the electrode has the form

$$I_i = -D_i \frac{dn_i}{dx} + n_i b_i E,$$

where D_i and b_i are the diffusion coefficient and mobility of ions, respectively; at high frequencies of the RF field and large ion masses the quantity **E** is the intensity of the quasi-stationary field in the layer.

In the experimental studies mentioned earlier, it is established that when the ions move in the NESCL, the ratio of the frequencies of the RF field f, the ion frequency $f_{ITF} = 1/\tau_{ITF}$ (τ_{ITF} – the time of flight by the ion in the NESCL) and the ionic plasma frequency f_{0i} is important. As can be seen in Figs. 2.2 and 2.3, the curves of the frequency dependence of the minimum RF voltage supporting the discharge and the RF voltage that sustains the discharge at a constant RF input power, reveal significant non-monotonicity in their course when the frequency f coincides with f_{ITF} and f_{0i}.

Earlier, an explanation was also given of the frequency dependence of the near-electrode jump in the potential $U_s(f)$ by the character of the ion transport to the electrode, presented in [145].

Essential information on the nature of the motion of ions in a collisionless NESCL is contained in a theoretical paper [210]. In it the motion of ions is described by the Mathieu equation

$$\frac{d^2x}{dz^2} - a_i(\in +2\cos 2z)x = 0,$$

where the parameter $a_i = \dfrac{4eV_0}{M_i d_s^2 \omega^2}$, $\omega = 2\pi f$, M_i is the mass of the

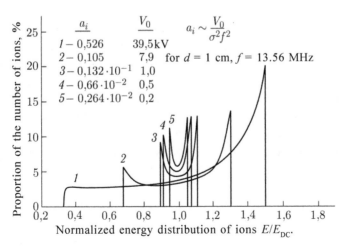

Fig. 2.26. The normalized energy distribution of ions for various the values of the parameter a_i $E_{DC} = eU_s$ [210].

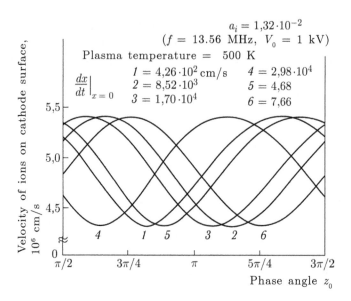

Fig. 2.27. Dependence of the ion velocity on the electrode on the phase angle of injection into the layer and the initial velocity for the parameter value a_i = 1.32 · 10^{-2} [210].

ion, the dimensionless parameter ϵ characterizes the stationary bias of the potential at the electrode, V_0 is the applied RF voltage, d_s is the thickness of the NESCL. The parameter a_i characterizes the physical conditions in which the ion moves. Thus, the ion energy spectrum obtained at the electrode must depend on the values of the quantities $V_\sim$, ω, d_s, M_i.

It was shown in [210] that for $a_i > 0.1$ the energy spectrum of ions arriving at the electrode should be wide, and the maximum ion energies $\varepsilon_{i\,max} > eU_s$, where U_s is the quasi-stationary voltage in the layer. The dependence of ε_i on the electrode on the phase angle of occurrence of the ion in the NESCL is calculated. The obtained energy spectrum of ions on the electrode for different values of the parameter a_i is shown in Fig. 2.26.

The dependence of the ion velocity on the electrode on the phase angle $\omega t = 2z_0$, at which the ion enters the NESCL with different velocities, is shown in Fig. 2.27.

An unexpected result of [210] is the appearance of ions on the electrode with energies $\varepsilon_{i\,max} > eU_s$. In this work, the following explanation is given: for large a_i, which means a large voltage amplitude $V_\sim$ or low frequencies ω or layer thickness d_s, the ions require a short transit time (about 3–4 field periods) to cross the layer. If the ion enters the NESCL at time t_1 and reaches the electrode at time t_2 (Fig. 2.27), it is accelerated by a voltage greater than U_s during the intervals Δt_1 and Δt_2 and therefore acquires an energy greater than eU_s.

For typical conditions of the original results presented in this monograph, the parameter a_i = 0.2–6.2. Thus, the energy interval of ions at the electrode was wide. More than 50% of the incoming ions had an energy $\varepsilon_i > eU_s$.

As a result, in the case of $U_s \geq 1$ kV, a sufficiently large number of ions could cause kinetic emission on the electrode of electrons with an emission factor $\gamma \geq 1$ el./ion [211]. From this it can be seen that in the RFCD it is possible, with the help of a change in the ion energy spectrum, to purposefully control the electron-emission processes at the electrodes.

2.10. Transport of electrons in the NESCL

2.10.1. On the presence of electrons in the NESCL

According to the traditional notions [23, 39, 212], the NESCL of the RFCD is free of electrons. Such assumptions are still being made. In some papers dealing with medium- and high-pressure RFD, it was assumed that the density of plasma electrons n_e falls off exponentially rapidly inside the NESCL in the transition region by an extent of the order of the Debye radius r_D [213].

Under such assumptions, the active conductivity of the NESCL is determined only by the motion of positive ions, and under certain regimes it is required to clarify the question of the mechanism for establishing the balance of charged particles in the stationary mode of RFCD.

Somewhat later, papers appeared that took into account the ingress of plasma electrons into the near-electrode region due to the periodic drop in the potential barrier near the electrode for them, in other words, due to the appearance of an 'anodic phase' at the electrode [138, 208, 210].

Meanwhile, in Refs. [51, 67] it has been repeatedly emphasized that in the NESCL of the RFCD the electrons are present during the entire period of the RF field, being in fluxes directed both from the plasma to the electrode, and from the electrode to the centre of the discharge gap.

This should lead to a significant increase in the active conductivity of the NESCL of the RFCD, which was later confirmed experimentally [74, 214].

One of the differences between the NESCL of the RFCD and the region of the cathode dip of the DCGD is that in the latter there are only electron fluxes from the cathode to the centre of the discharge.

As the mechanisms for the entry of electrons into the NESCL of the RFCD, we note the following: 1) the emission of electrons from the surface of the electrodes; 2) the arrival of the main mass of plasma electrons in the NESCL due to a periodic decrease in the height of the near-electrode potential barrier; 3) the proposed new, discrete mechanism of transport of plasma electrons to the electrodes, described below. The results presented below show that the presence of electrons in the NESCL is an essential moment in the mechanism of maintaining RFCD.

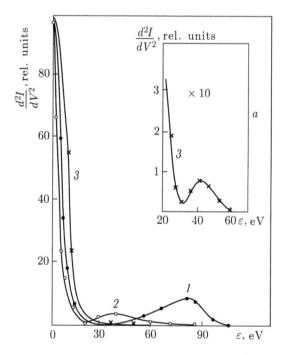

Fig. 2.28. Dependence of the nature of the energy distribution of electrons from the direction in space. He, $p = 0.5$ Torr. Near-electrode region, electrode–probe distance $d_{ep} = 10$ mm, $f = 8$ MHz, $V_\sim = 250$ V. *1* – the surface of the plane probe faces the electrode, *2* – the plane of the probe is parallel to the axis of the discharge tube, *3* – probe facing the electrode of the vitrified side.

2.10.2. Experimental study of near-electrode electron fluxes from electrodes to plasma

The motion of electrons in the direction from the electrode to the interior of the discharge gap was studied using a probe and a number of optical methods.

The spatial nature f the motion of electrons in the near-electrode region was studied by elucidating the presence of anisotropy of the probe characteristics of a plane oriented probe and the corresponding electron distribution functions with respect to the energy $f_e(\varepsilon)$.

The function $f_e(\varepsilon)$ was determined by automatically measuring the second derivative of the probe characteristic d^2I_3/dV_3^2 using a standard electronic circuit for such measurements.

The qualitative nature of the energy distribution of electrons, obtained for different orientations in the space of a plane probe in the near-electrode region of the discharge is shown in Fig. 2.28. The curves d^2I_3/dV_3^2 are presented instead of the distribution functions, since in the case of anisotropic plasma it is incorrect to construct the function $f_e(\varepsilon)$ from the known second-derivative curve [215].

As can be seen from the results obtained (Fig. 2.28, curves *1*, *2*, *3*), the electron energy distribution in the vicinity of the electrode essentially depends on the direction in space.

When the plane probe is located by its 'face' to the electrode (curve *1*), the curve $d^2I(V)/dV^2$ in the region of high electron energies ($\varepsilon_e \sim 50$–100 eV) has an additional maximum. If the working surface of the probe was located perpendicular to the surface of the electrode, the additional maximum on curve *2* became much smaller and shifted

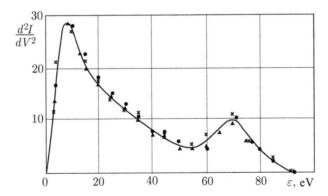

Fig. 2.29. Dependence of the nature of the energy distribution of electrons from the direction in space. He, $p = 0.5$ Torr, $f = 7$ MHz, $V_\sim = 360$ V, $d_{as} = 25$ mm. The notation for the experimental points is the same as in Fig. 2.28.

to the region of lower energies ($\varepsilon_e \sim 25$–70 eV). When the probe was oriented with the vitrified side to the electrode, curve *3* smoothly dropped to zero, and only a tenfold increase in the sensitivity of the measurement circuit made it possible to detect an additional small maximum in the region $\varepsilon_e \sim 30$–60 eV (Fig. 2.28).

Studies have shown that under these conditions at distances of the order of 20 mm from the electrode the anisotropy of the electron energy distribution disappeared (Fig. 2.29).

Thus, the results presented above show that in the RFCD studied there are anisotropic electron fluxes – electron beams directed from the electrode into the interior of the discharge gap.

This experimentally observed phenomenon of occurrence in the RFCD of high-energy near-electrode electron beams was presented at the Second All-Union Conference on the Physics of Low-Temperature Plasma in Minsk in 1968 [47, 48] and had far-reaching consequences for studying the physics of high-frequency capacitive discharges of low and medium pressure.

Important additional information on the nature of the motion of electrons in the near-electrode regions of RFCD was obtained by optical methods, some of which were used to study DCGD [182], which has much in common with the investigated RF discharge.

In this case, studies were carried out: 1) the dependence of the intensity of quasi-stationary luminescence on the distance to the electrode $I(x)$; 2) distributions $I(x)$ in one discharge tube for RFCD and DCGD; 3) the dynamics of the distribution structure $I(x)$ during the period of the RF field; 4) the dependence of the character of the local luminescence oscillations $I(t, x)$ on the distance to the electrode; 5) the dependence of the amplitudes of the luminescence oscillations on the frequency of the RF field and its second harmonic on the distance to the electrode – $I_f(x)$ and $I_{2f}(x)$.

The experimental system for studying the optical radiation of RFCD which had two channels: 1) to study the spatial distribution of the intensity of a stationary glow; 2) to study the oscillations in time of the luminescence intensity, was given earlier in Fig. 1.54.

The sources of RF power included PSK-4, UVCh-66, Brig and others were used in the frequency range 0.5–25 MHz, with a fixed frequency of 40 MHz.

The amplitude of RF voltage was measured by voltmeters such as VK 7-9, V 7-15 and VU-15 with a capacitive voltage divider. The frequency of RF voltage was constantly monitored with the help of the Ch3-34 frequency counter and the shape – with the help of the S1-31 oscilloscope.

The integral luminescence of RFCD was investigated using a photomultiplier FEU-79 with a high sensitivity in the visible spectral range.

In addition to the photomultiplier, the first channel of the system included an I-37 direct current amplifier and an N-37 recorder, and the second channel – a broadband amplifier with a bandwidth of 30 MHz and an amplification factor of $K = 100$ and an oscilloscope of type S1-31.

When studying the dynamics of the spatial structure of the luminescence during the period of the RF field, an optical signal through the diaphragm system or from the output slit of the MDR-12 monochromator came to the FEU-38, whose output was connected to the S7-13 strobe oscilloscope. The analog output of this oscilloscope was connected to a two-coordinate recorder LKD-4.

In these measurements, the frequency of the field was chosen sufficiently low ($f = 2.4$ MHz), so that the relation $\tau_{ex} < 1/f$ for the most intense He lines, where τ_{ex} is the lifetime of the atom in the excited state, was fulfilled.

In the stroboscopic mode of measurement the strict periodicity of the signal was monitored under the experimental conditions. In this case, two types of measurements were made: obtaining 'phase' and 'time' characteristics of the RFCD emission.

To obtain the 'phase' characteristics, the S7-13 oscilloscope strobe was installed in series in four characteristic phases of the RF field period, and the discharge tube was smoothly moved before the input of the diaphragm system by means of a movable platform with an electric motor.

When recording the 'time' characteristics, the position of the tube was fixed, and the strobe moved smoothly within 1–2 periods of the RF field. These characteristics were obtained for different distances from the electrode.

With the accuracy to a constant component of the luminescence intensity, the results obtained by the two methods mentioned above must coincide, which was confirmed in the measurements.

The amplitudes of the luminescence intensity oscillations at the frequency of the RF field and its second harmonic were studied using a selective microvoltmeter of the V6-10 type, and a spectral analyzer of the SK 4-58 type was used to visually observe the relationship between the harmonics.

The spatial resolution during the described measurements was 0.1–0.5 mm.

Special measures were taken to exclude the effect of processes on one electrode on the near-electrode physical conditions of another.

Let us consider the experimental results obtained.

Earlier, Fig. 1.35 showed the recorded change in the spatial distribution of the luminescence intensity in the near-electrode region of the RFCD with a monotonic increase in the applied RF voltage. At the same time, increasing RF voltage resulted in the formation of the luminescence distribution $I(x)$ which is identical to that in DCGD. A typical distribution of this type, characteristic of the γ-type RFCD, is shown in Fig. 1.33. Here all the spatial regions of the DCGD are already present – an Aston dark space, a cathode glow, a Crookes dark space, a negative glow, a Faraday dark space, and a positive column common for the two near-electrode regions.

Identical distributions of the integral luminescence $I(x)$ in the near-electrode region of the RFCD and the cathode region of the DCGD were obtained in the same discharge tube by alternately igniting these discharges (Fig. 1.34).

The distributions $I_\lambda(x)$ of the intensity of the spectral line He I 5876 Å in the corresponding regions of the DFCD and DCGD are shown in Fig. 2.30.

The qualitative similarity between the quasi-stationary distributions of $I(x)$ and $I_\lambda(x)$ in RFCD and DCGD is obvious.

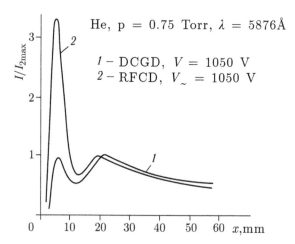

Fig. 2.30. The intensity distributions of the He I 5876 Å spectral line normalized to the 2nd maximum of curve *2* in the near-electrode region of the RFCD and the cathode region of the DCGD in a single discharge tube.

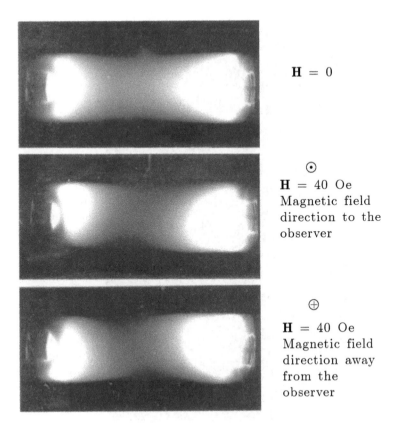

H = 0

⊙
H = 40 Oe
Magnetic field
direction to the
observer

⊕
H = 40 Oe
Magnetic field
direction away
from the
observer

Fig. 2.31. Change in the luminescence pattern of the RFCD under the action of a transverse homogeneous magnetic field **H**. RFCD. Ne. p = 0.3 Torr, $V_\sim$ = 2000 V, f = 1.2 MHz (active electrode on the right).

Since the motion of electrons in the cathode region of the DCGD is unidirectional, from the cathode to the centre of the discharge, then, based on what was said above, in the near-electrode regions of the RFCD one should also assume that the emitted electrons are not vibrational but unidirectional.

A direct proof of this is the picture observed when a constant uniform magnetic field $\mathbf{H}_0$ is applied perpendicular to the axis of the discharge tube (Fig. 2.31). It is seen that the glow of the electrodes deviates in different directions, since the near-electrode electron fluxes are directed towards each other. When the direction of the field $\mathbf{H}_0$ reverses, the direction of the emission deviation at the electrodes changes in a similar manner.

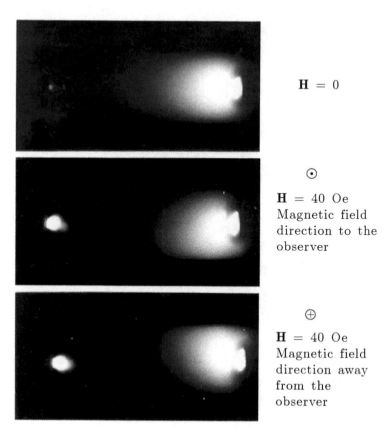

H = 0

⊙

H = 40 Oe
Magnetic field
direction to the
observer

⊕

H = 40 Oe
Magnetic field
direction away
from the
observer

Fig. 2.32. Change in the luminescence pattern of a DC~GD under the influence of a transverse homogeneous magnetic field **H**. Ne. $p = 0.3$ Torr, $V_{ak} = 2000$ V, $I_p = 0.4$ mA (cathode on the right).

When a magnetic field is applied to the discharge interval of the DCGD, the glow at the cathode deviates in the same way as the glow of the near-electrode region of the RFCD (Fig. 2.32).

It is well known that the directional motion of electrons in the cathode region of the DCGD causes polarization of the radiation in this discharge region [182].

To study the polarization properties of RFCD radiation, the highly sensitive polarization spectroscopy technique developed at the University of Leningrad was used, based on the Hanle effect [216, 217].

In the experiments carried out jointly with the University of Leningrad staff of the S.A. Kazantsev group, RFCD in inert gases (He, Ne, Ar) was excited in a discharge tube 4 cm in diameter and 6 cm in length, placed between the outer planar electrodes.

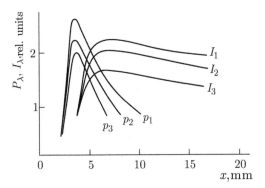

Fig. 2.33. The dependence of the intensity of the spectral line Ar I 6032 Å and its polarization along the discharge gap of the RFCD. Ar, $p = 0.023$ Torr, $f = 100$ MHz.

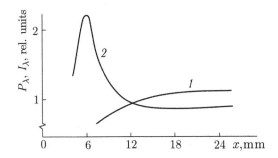

Fig. 2.34. Dependence of the intensity of the spectral line He I 4922 Å (*1*) and its polarization (*2*) along the discharge gap of RFCD. He, $p = 0.02$ Torr, $V_\sim = 100$ V, $f = 100$ MHz.

Under the investigated conditions, neutral atoms were excited by collisions of the first kind with electrons. Consequently, the degree of linear polarization of the spontaneous emission is uniquely related to the quadrupole moment of the electron velocity distribution function $f_e(v)$, due to the anisotropy of the electron motion [217].

The distribution of the degree of polarization $P_\lambda(x)$ of the radiation of a number of spectral lines along the axis of the discharge tube was investigated using a magnetic polarization spectrometer [216]. At the same time, the intensity distribution of the spectral lines $I_\lambda(x)$ was studied.

For all the investigated spectral lines in the near-electrode region, a maximum was recorded for the $P_\lambda(x)$ curve, which was located closer to the electrode than the maximum radiation intensity (Fig. 2.33).

At the lowest investigated pressures, when the curve $I_\lambda(x)$ no longer exhibited a near-electrode maximum, the near-electrode polarization maximum was clearly observed (Fig. 2.34).

Thus, the fixed polarization of the radiation in the near-electrode regions of RFCD also indicates the presence of directed electron fluxes there, and the course of the $P_\lambda(x)$ curve demonstrates the evolution of these fluxes in space, up to their complete damping.

In the process of work, ultrafast photography of the near-electrode regions of the RFCD was also performed with the help of the 'time magnifier' LV-03. The observed character of the glow behaviour in time and space is shown in Fig. 2.22. Hence it is obvious that the glow in the vicinity of the electrode is of a pulsed nature, propagating from the electrode deep into the discharge gap during the negative half-cycle of the RF field. This additionally indicates the existence of pulsed electron fluxes in a given discharge, confirming the hypothesis advanced earlier in this paper on the formation in the near-electrode region of pulsed constant fields accelerating the electrons emitted by the electrode.

The data of stroboscopic studies of the change in the luminescence structure in the near-electrode regions ('phase' and 'time' characteristics) during the RF field period will be considered below in connection with a detailed study of the dynamics of the near-electrode electron beams.

The results of these studies of the nature of the motion of electrons in the near-electrode regions of RFCD were also experimentally confirmed in foreign works [218, 219]. Thus, in particular, in [219], by photographing the near-electrode regions of RFCD with cellophane screens installed inside them, regions of the geometric shadow that are opaque to electron currents were fixed in the corresponding half-periods of the RF field. These authors came to the conclusion that electrons from the electrode move along the 'ballistic' trajectories deep into the discharge, in the vicinity of the electrode.

2.10.3. Transfer of electrons from the plasma to the electrodes

There are a number of points that are of interest to the study of the transport of plasma electrons to RFCD electrodes.

Investigation of the corresponding electron fluxes makes it possible to determine the electron energetic spectrum (EES) of a near-electrode plasma, to clarify the mechanism for ensuring the balance of charges in the discharge, and to develop an effective measurement technique with the help of an energy analyzer of charged particles.

For most of the duration of the high-frequency period, plasma electrons are held by a near-electrode potential barrier of variable height, reaching a maximum value in the negative half-cycle of the RF field and a minimum in the positive one.

When the potential barrier decreases, the plasma begins to diffuse to the electrode, and the NLB disappears if the bulk of the 'slow electrons' reaches the electrode.

Let us determine the characteristic distances to which the plasma is displaced to the electrode, starting from the maximum extension of the NESCL, for the two modes of the discharge studied, which differ sharply from the average energies of the groups of 'slow' electrons: 1) symmetrical RFCD; 2) asymmetric RFCD.

Estimates are made for the characteristic physical conditions: RFCD in He, $p = 0.1$ Torr, $d_s \sim 1$ cm, $f = 1$ MHz, $T_i = 0.05$ eV.

The length of diffusion displacement x of plasma over time $\tau = T/2$ is equal to $x \sim (D_a \tau)^{1/2}$, where $D_a \approx \dfrac{kT_e b_i}{e}$ is the ambipolar diffusion coefficient, $b_i = \dfrac{e}{M_i \nu_{im}} = \dfrac{e}{M_i n_a \sigma_t \nu_{iT}}$ is the mobility of the ions, $n_a = \dfrac{p}{kT_a}$ is the concentration of the atoms, σ_t is the recharge cross section of the ions, T_a is gas temperature, $\nu_{iT} = \left(\dfrac{3kT_i}{M_i} \right)^{1/2}$ is the thermal velocity of the ions. We have:

$$\tau = \frac{T}{2} = 5 \cdot 10^{-7} \text{ s}, \quad T_i = 500 \text{ K}, \quad M_i = 6.64 \cdot 10^{-24} \text{ g, according to [157]:}$$

$$\sigma_t = 0.283 \cdot 10^{-16}, \quad P_t p = 0.283 \cdot 10^{-16} \cdot 100 \cdot 0.1 = 2.8 \cdot 10^{-16} \text{ cm}^2.$$

$$D_a = \frac{T_e}{n_a \sigma_t} \left(\frac{k}{3M_i T_i} \right)^{1/2}, \quad n_a = \frac{0.1 \cdot 1333}{1.38 \cdot 10^{-16} \cdot 300} = 3 \cdot 10^{15} \text{ cm}^{-3}.$$

Consequently, we obtain:
  1. Symmetric RFCD

$$T_e = 46 \cdot 10^3 \text{K},$$

$$D_a = \frac{46 \cdot 10^3}{3 \cdot 10^{15} \cdot 2.8 \ 10^{-16}} \left(\frac{1.38 \cdot 10^{-16}}{3 \cdot 6.64 \cdot 10^{-24} \cdot 500} \right)^{1/2} = 6.5 \cdot 10^6 \text{ cm}^2 \cdot \text{s}^{-1},$$

$$x = (D_a \tau)^{1/2} = (6.5 \cdot 10^6 \cdot 5 \cdot 10^{-7})^{1/2} = 1.8 \text{ cm}.$$

From this one can see that $x > d_s$, so that the plasma travels to the electrode and should completely disappear.

2. Asymmetric RFCD

$$T_e \approx 10^3 \text{ K}; \quad D_a = \frac{10^3}{3 \cdot 10^{15} \cdot 2.8 \cdot 10^{-16}} \left(\frac{1.38 \cdot 10^{-16}}{3 \cdot 6.64 \cdot 10^{-24} \cdot 500} \right)^{1/2} =$$

$$= 1.4 \cdot 10^5 \text{ cm}^2 \cdot \text{s}^{-1}, \quad x = (D_a \tau)^{1/2} = \left(1.4 \cdot 10^5 \cdot 5 \cdot 10^{-7} \right)^{1/2} = 0.25 \text{ cm}.$$

Thus, $x < d_s$, and for a half-period of the field the plasma has little time to shift to the electrode. In this case, only the fastest electrons from the velocity distribution function can reach the electrode. Their concentration is significantly lower than the concentration of ions in the NESCL, therefore, passing through the layer, they practically do not change the parameters. In this case, the plasma does not touch the electrode during the entire period of the RF field.

The tendency of the NESCL to be stationary increases with increasing RF voltage $V_\sim$ and frequency f of the field. As follows from the experimental results given below, with increasing $V_\sim$ the effective temperature $T_{e \text{ eff}}$ of the main mass of electrons decreases and, accordingly, the coefficient of their diffusion decreases.

The optical observations carried out in the present work showed that with increasing $V_\sim$ the degree of stationarity of the NESCL parameters increases, which is expressed in a decrease in the amplitude of the variable component of the luminescence intensity in the near-electrode regions of the RFCD.

Thus, under the investigated conditions, the NESCL of the RFCD could both disappear for a short time (to experience complete collapse) and to exhibit a different degree of stationarity of its parameters.

The main regularities of the transfer of electrons and ions to the electrodes of a symmetrical RFCD, one of which is grounded, can be conveniently analyzed qualitatively with the help of plots of the plasma potential V_p and the active electrode V during the period of

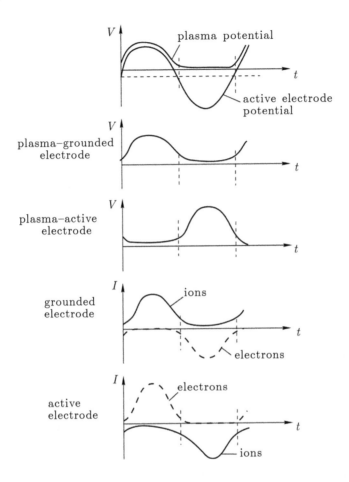

Fig. 2.35. Voltage and charge flux in the NESCL during the RF field period. f = 380 kHz [138].

the RF field (Fig. 2.35), given in [138]. The analysis is carried out for low (f = 380 kHz) and relatively high (f = 13 MHz) electric field frequencies.

The qualitative course of the fluxes of charged particles in a time and the electrodes in the case of a low frequency is shown in Fig. 2.35 and is obvious without any special comments.

At an increased frequency of the RF field, a quasi-stationary potential difference U_s occurs between the plasma and the grounded electrode (Figure 2.36). This potential barrier for plasma electrons is much lower in the positive half-cycle of the RF field and is noticeably higher for the negative half-period, compared with the RFCD at f = 380 kHz.

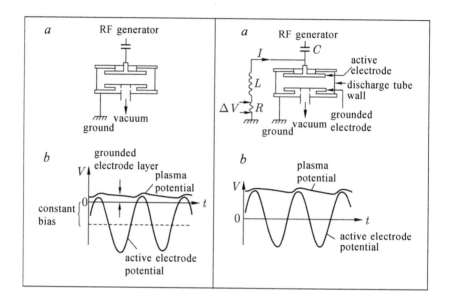

Fig. 2.36. Connection diagrams: *a*) without DC grounding and with DC grounding and *b*) the corresponding view of the potential of the active electrode and the plasma potential [138].

The main difference for the electron fluxes in the negative half-cycle for low and high frequencies is that in the latter case, the slowest electrons of the EES of the near-electrode plasma do not enter the electrode.

Obviously, for earlier experimental results with frequency $f = 1.2$ MHz, an intermediate case with a relatively low potential barrier for electrons in front of the grounded electrode was observed.

It is emphasized that the above was a question of RFCD in the V_0-mode. According to the results of [138], when connecting an RF generator to a discharge gap without a transient capacitance (I_0-mode), a significant constant potential difference arises before the ground electrode, which is completely unacceptable for the studies considered in this section, in particular, due to the use of an energy analyzer on a grounded electrode.

The main differences in the nature of the transport of charged particles to the electrodes of symmetric and asymmetric RFCDs can be qualitatively illustrated on the basis of the curves of the variation of the potential difference ΔV in the NESCL of both electrodes during the RF field, presented in [208] (Fig. 237).

First of all, the situation in the NESCL of a grounded electrode is of interest.

As can be seen from Fig. 2.37, in the case of a symmetrical RFCD, the value of ΔV varies from zero to the amplitude value of the RF voltage in the NESCL of both electrodes. In this case, the processes of charge transfer to the electrodes and the formation of near-electrode electron beams will be identical for the above-mentioned electrodes.

According to the same Fig. 2.37, in asymmetric RFCD there should be a noticeable quantitative difference in the parameters of the corresponding processes in the NESCL of the opposing electrodes.

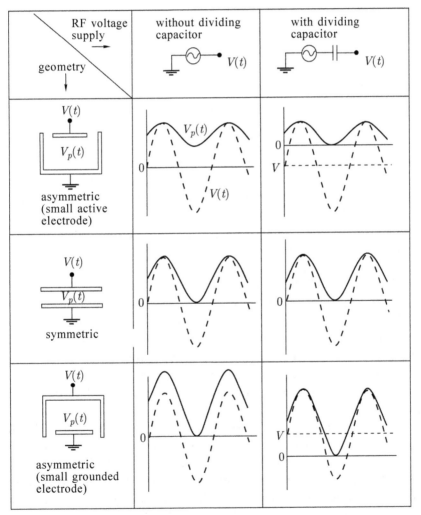

Fig. 2.37. The course of the plasma potential for different variants of the RF voltage supply and different discharge geometries [208].

Indeed, in the present work, ARFCD was investigated with a grounded electrode with a diameter $D_g = 5.8$ cm and an active electrode with $D_a = 0.5$ cm. Thus, the ratio of the electrode areas of this RFCD was

$$\frac{S_g}{S_a} = \frac{D_g^2}{D_a^2} \cong 135.$$

It follows that, according to the formulas (2.9), (2.10) [78], the ratio of the energies of the near-electrode electron beams of the active ε_{eb}^a and the grounded ε_{eb}^g the electrodes should be

$$\frac{\varepsilon_{eb}^a}{\varepsilon_{eb}^g} = \left(\frac{S_g}{S_a}\right)^q \approx 135^{1.5} = 1.6 \cdot 10^3,$$

The density of these beams should also differ significantly

$$\frac{n_{eb}^a}{n_{eb}^g} \sim \frac{S_g}{S_a} > 10^2.$$

Thus, the interaction of the beams from the active electrode to the grounded electrode seems to be quite intense for its experimental study, in contrast to the interaction of the beams from the grounded electrode with the active electrode. Therefore, it was considered possible to fix the signal from the active electrode beam using an energy analyzer on the grounded electrode.

In the case of the investigated symmetrical RFCD with electrodes of diameter $D = 3.2$ cm, the areas of the latter exceeded those for the active electrode ARFCD by more than 40 times. Consequently, both the n_{eb} and ε_{eb} energies of the near-electrode beams in the symmetric RFCD were significantly smaller than the corresponding characteristics of the active electrode beams ARFCD (n_{eb}^a, ε_{eb}^a). Therefore, it was unlikely to fix the beam signal of the active electrode with an energy analyzer on a grounded electrode in the examined symmetrical RFCD.

In these studies, the charged particle fluxes on the grounded electrode in RFCD were studied by oscillographing the signal from the load resistance of the energy analyzer described below. At the same time, the energy analyzer was switched on in the screened collector mode, i.e., when all grids of the energy analyzer, except for closing the inlet, were short-circuited with the collector. In these experiments, the analyzer did not contain any analyzing electrical voltages, and the charged particle fluxes flowed freely from the discharge gap to the collector. So the recorded signal corresponded to the total flux of electrons and ions to the collector I_c, representing the difference signal of the above flows.

In spite of what was said above, with the help of this technique quite satisfactory information was obtained on the time behaviour of the ion and electron fluxes separately. This was due to the fact that the amplitudes and time course during the period of the RF field of the ion and electron fluxes, as can be seen from the oscillograms $I_c(t)$ given below, differed significantly.

Typical oscillograms $I_c(t)$, obtained in experiments with symmetric and asymmetric RFCD under different physical conditions, are shown in Fig. 2.38, 2.39.

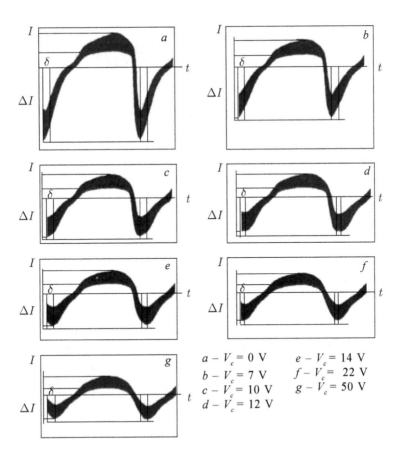

$$
\begin{array}{ll}
a - V_c = 0 \text{ V} & e - V_c = 14 \text{ V} \\
b - V_c = 7 \text{ V} & f - V_c = 22 \text{ V} \\
c - V_c = 10 \text{ V} & g - V_c = 50 \text{ V} \\
d - V_c = 12 \text{ V}
\end{array}
$$

Fig. 2.38. The oscillograms of the fluxes of charged particles to the electrode (air, $p = 0.15$ Torr, $V_\sim = 700$ V, $f = 1$ MHz) for various potentials V_c at the collector of energy analyzer.

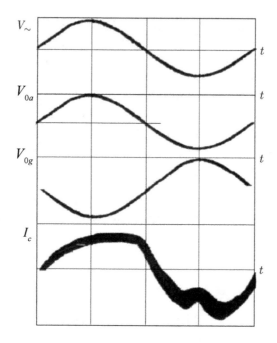

Fig. 2.39. The signal from the collector I_c of the energy analyzer, as well as $V_\sim$, V_{0a}, V_{0g}, phased over the period of the RF field.

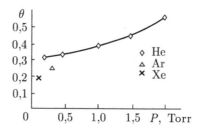

Fig. 2.40. The dependence of the parameter $\Theta = t_e/T$ of the electron pulse on the pressure and kind of gas.

The above oscillograms show that a relatively weakly modulated flux of positive ions passes through the electrode during the entire period of the RF field T, and the electron fluxes are of an impulse nature. The duration of these electron pulses t_e can be sufficiently small $t_e \ll T$. For the time characterization of these pulses, we introduce the parameter $\Theta = t_e/T$.

The dependence of Θ on the pressure and the type of gas obtained experimentally is shown in Fig. 2.40.

The time course of the electron fluxes to the electrode essentially depends on the RFCD parameters.

It was also found when applying voltages V_k retarding the electrons to the collector of the energy analyzer that the electronic pulses change their form for different V_k (Fig. 2.38). This indicates that during the electron pulse there are time-varying fluxes of electrons of different energies. Such situations will be discussed below.

The received oscillograms $I_k(t)$ show that for each period of the RF field at the electrode, a balance of positive and negative charges is formed provided by the corresponding value of the parameter Θ of the electron pulse.

As can be seen from Fig. 2.40, the fraction of the electron pulse to electrode in the duration of the RF field period $\Theta = t_e/T$ increases with increasing gas pressure and decreases with increasing atomic weight of the working gas particles. This is to be expected, since the electron pulse compensates for the positive charge brought by the ions to the electrode over the entire period of the RF field. As the ion mass grows, the charge they bring on the electrode decreases over the period of the field, so the electrons have time to compensate it for a shorter pulse duration.

Let us briefly discuss the nature of the motion of charged particles through the NESCL from the plasma to the electrode.

Under the experimental conditions, according to the estimates and experimental data, the thickness of the near-electrode layer is $d_s \sim 1$ cm.

Under the experimental conditions (He, air, $10^{-2} < p \le 10^{-1}$ Torr), the ion motion regime can be considered to be slightly collisional, since the mean free path of ions is $\lambda_i \le d_s$.

The motion of electrons in the near-electrode layer under the conditions under consideration can be considered collisionless; practically at any electron energies $\lambda_e > d_s$.

We note some peculiarities of the time course of physical processes in the NESCL of the RFCD. According to literature data [138, 208], the RFCD electrodes either permanently have a potential V below the plasma potential V_p, or exceed the latter for a time Δt much shorter than the period of the RF field ($\Delta t \ll T$) [207]. In this connection, the potential barrier of variable height for plasma electrons exists continuously or disappears only for a time $\Delta t \ll T$. A number of experimental studies show that such a quasi-stationary NESCL exists [67, 69].

The subject of our consideration is the near-electrode plasma created by a combination of physical processes in the NESCLof the RFCD. Here it should be noted that in the symmetrical and asymmetrical RFCDs studied, the physical conditions were significantly different. In the first case, the plasma was created by processes that occurred at the electrode adjacent to the plasma, and in the second case, mainly by processes that occurred at the opposite (active) electrode.

Naturally, the physical processes in the near-electrode plasma and the NESCL are self-consistently interrelated. Thus, for example, in this paper, the relationship between EES plasma and the character of the electron transfer from the plasma to the electrode has been observed [220].

Figure 2.24 *a, b* shows the oscillograms of the charge flows $I_c(t)$ on the grounded electrode of a symmetric RFCD in He and Xe, respectively.

From this it can be seen that in the case of RFCD in Xe, the leading edge of the electron pulse to the electrode is very sharp (practically vertical), and the function of the main electron mass distribution $f_e(\varepsilon)$ determined using the energy analyzer occupies a narrow energy interval [220].

In the RFCD in He, the leading edge of the analogous electron pulse is stretched in time, and the distribution function $f_e(\varepsilon)$ of the slow plasma electrons, which are predominant in the number, as is shown below, is characterised by the energy gap broader than in Xe.

Thus, there is a fairly strong correlation of the EES of the near-electrode plasma with the nature of the electron transport on the electrode. This experimental fact can be explained as follows.

The investigated RFCDs were supported by sufficiently high RF voltages $V_\sim \sim 1$ kV. It follows that the EES of the near-electrode plasma must be wide and contain electrons with energies $0 < \varepsilon_e \le 1$ keV. However, due to the lack of sensitivity of the energy analyzer in conditions of a symmetrical RFCD with a relatively large area of electrodes, relatively small high-energy electrons were not recorded.

In the case of RFCD modes with high values of $V_\sim$ and f, high-energy near-electrode electron beams are created, creating a near-electrode plasma of the 'negative glow' (NG) plasma type. The negative glow plasma, as is known [151], has an extremely low electron temperature $T_e \sim 10^{-3}$–10^{-1} eV. As a result, the diffusion coefficient of electrons $D_e \sim T_e$ must also be very small. Therefore, in a short time of the 'anode' phase of the electrode, the electrons will have not time to reach the electrode. The question arises of the mechanism of the transport of plasma electrons to the electrodes, which provides a balance of charges of the opposite sign in the discharge gap of the stationary RFCD. The solution of this problem is proposed in the next chapter in the form of a new, discrete mechanism of electron transport to the RFCD electrodes.

Thus, depending on the physical parameters of the RFCD, the transport of plasma electrons to the electrodes providing a quasi-stationary discharge state can occur by two mechanisms: 1) traditional, continuous – by diffusion of the plasma to the electrode during the positive half-cycle of the RF field, or 2) the proposed new, discrete mechanism of electron transfer.

Above we talked about the transport of electrons to the active electrode. On a grounded electrode, in addition to the diffusion flux of thermal electrons, a beam of high-energy electrons from the active electrode enters the positive half-cycle of the RF field.

In order to complete the picture of the transport of electrons to the electrodes of the RFCD, we recall that in the theoretical paper [78] it was predicted that in the ARFCD (I_0-mode) the electrons leave only through the active small area electrode, being locked by a constant potential barrier near the grounded electrode. This conclusion was confirmed experimentally in [145].

The material presented in this section shows that the transfer of plasma electrons to the electrodes in all the investigated modes of RFCD takes place, and the mechanism of this process is ambiguous.

The determining mechanism for maintaining RFCD depends on the physical nature of the total discharge current $I_\sim$ passing through the NESCL – on the ratio of its active and reactive components.

In its properties, RFCD is closer to DCGD tthe larger the contribution to the total current $I_\sim$ is made by the conduction current, determined by the motion of charges on the electrodes.

In many of the experiments discussed here, we used lower frequencies of the field $f \geq 1$ MHz, which significantly reduced the contribution of the bias current to the total discharge current.

The material presented in this section shows that the processes of transport of charged particles to the electrodes play an important role in the physical mechanism of RFCD.

It was noted earlier that, being the most inert, conservative component of the plasma, the ions ensure the tendency of the NESCL to be stationary, move in quasi-stationary fields.

The transport of electrons to the electrodes self-adapts to the ionic one, providing a quasi-stationary state of the balance of charged particles in the discharge. Moreover, in the RFCD modes with the absence of the NESCL collapse, short-duration pulsed low-density electron fluxes to the electrodes relatively slightly modulate the parameters of the NESCL.

2.11. Electron-emission processes at electrodes

In the fifties of the last century, it was suggested to distinguish between two types of low- and medium-pressure RF discharges – α- and γ-discharges [23]. It was assumed that in the γ-discharge, electron-emission processes at the electrodes play a decisive role in its physical mechanism. However, in the literature there is practically no information on experimental studies of emission processes at RFCD electrodes.

In the works of the author and co-workers [94, 145, 221], an original spectroscopic technique for measuring the electron density n_{e0} emitted from the surface of the RFCD electrodes was proposed and applied. The essence of this technique is as follows.

It was previously experimentally established that the spatial structure of the luminescence of the RFCD of the γ-type in detail reproduces the distribution of the luminescence in the discharge gap of the DCGD. At the same time, the key point is the well-known position on the nature of the excitation of radiation emitted from the cathode by electrons in the cathode emission of DCGD [182].

The intensity of cathode emission from the cathode to its maximum is proportional to the density of the electrons emitted from the cathode surface, for which the processes of avalanche propagation have not started up to the indicated location. Assuming that at the location of the maximum intensity of the cathode luminescence (CL), electrons acquire an energy $\varepsilon_{e\,max}$ corresponding to the maximum excitation cross section of a certain spectral line [182], the expression for the absolute intensity of the spectral line has the form:

$$I_{ki} = n_e^{CL} n_a \sigma_{k\,max} v_{e\,max} \frac{A_{ki}}{\sum_r A_{kr}} h\nu_{ki},$$

(2.12)

where n_e^{CL} is the electron concentration at the maximum of the CL region; n_a is the concentration of neutral atoms; $\sigma_{k\,max}$ is the maximum value of the excitation cross section for the k-th level of the atom by direct electron impact; $v_{e\,max}$ is the electron velocity corresponding to the maximum value $\sigma_{k\,max}$; A_{ki} is the probability of a spontaneous transition. $k \rightarrow i$; $\sum_r A_{kr}$ is the sum of the probabilitiesof spontaneous transitions from the k-th level to all lower levels r; hv_{ki} is the energy of the radiation quantum with frequency v_{ki}.

Hence, by measuring the absolute emission intensity of the spectral line I_{ki}, we can determine the electron density n_e^{CL}:

$$n_e^{CL} = \frac{I_{ki}}{n_a \sigma_{k\,max} v_{e\,max} hv_{ki} A_{ki} / \sum_r A_{kr}}. \tag{2.13}$$

Let us now find the density of electrons emitted from the electrode surface n_{e0}. Assuming that the flux of emitted electrons does not ionize and do not dissipate up to the CL region, we can write the expression for the flux density of the emitted electrons:

$$j_{e0} = n_{e0} v_{e0} = n_e^{CL} v_{e\,max}, \tag{2.14}$$

where v_{e0} is the average exit velocity of electrons emitted from the surface of the electrode. Hence, using the expression (2.14), for the density of the emitted electrons n_{e0}, we have:

$$n_{e0} = n_e^{CL} \frac{v_{e\,max}}{v_{e0}} = \frac{I_{ki}}{n_a \sigma_{k\,max} v_{e0} hv_{ki}} \cdot \frac{\sum_r A_{kr}}{A_{ki}}. \tag{2.15}$$

The value n_{e0} can also be determined by measuring the relative intensity of the spectral line I_{ki} in the cathode luminescence I_{ki}^{CL} and the positive column I_{ki}^{nc} of the discharge, if any method is used to determine n_e^{nc} and T_e^{nc} in a positive column plasma. Then can be written:

$$I_{ki}^{CL} = n_e^{CL} n_a \sigma_{k\,max} v_{e\,max} \frac{A_{ki}}{\sum_r A_{kr}} hv_{ki}, \tag{2.16}$$

$$I_{ki}^{nc} = n_e^{nc} n_a \langle \sigma_k(v) v_e \rangle \frac{A_{ki}}{\sum_r A_{kr}} hv_{ki}. \tag{2.17}$$

According to (2.16) and (2.17) we get

$$\frac{I_{ki}^{CL}}{I_{ki}^{nc}} = \frac{n_e^{CL} \sigma_{k\,max} v_{e\,max}}{n_e^{nc} \langle \sigma_k(v) v_e \rangle}$$

and then

$$n_e^{CL} = n_e^{nc} \frac{I_{ki}^{CL}}{I_{ki}^{nc}} \cdot \frac{\langle \sigma_k(v)v_e \rangle}{\sigma_{k\,max} v_{e0}}. \tag{2.18}$$

Using equation (2.14), gives

$$n_{e0} = n_e^{CL} \frac{v_{e\,max}}{v_{e0}} = n_e^{nc} \frac{I_{ki}^{CL}}{I_{ki}^{nc}} \frac{\langle \sigma_k(v)v_e \rangle}{\sigma_{k\,max} v_{e\,max}}. \tag{2.19}$$

For the development of the measurement methodology, measurements were also taken in DCGD. In this case, the unknown quantity n_{e0} can be determined additionally as follows:

$$i = jS_{cat}, \tag{2.20}$$

where i is the discharge current; S_{cat} – cathode area; $i = i_e + i_i = i_i(1 + \gamma)$, i_i is the ion current to the cathode; i_e – current of emission of electrons from the cathode; $j_e = \gamma j_i$ – electron emission current density

$$j_e = \gamma \frac{i}{(1+\gamma)S_{cat}} = n_{e0}v_{e0}, \tag{2.21}$$

from which $n_{e0} = \frac{\gamma}{1+\gamma} \cdot \frac{i}{S_{cat}v_{e0}}$, where the numerical values of γ and v_{e0} are taken from the literature.

In methodical measurements in a DC discharge in He, the value of n_{e0} was determined by the three indicated methods, and in the RF discharge with external electrodes only the first. Helium was chosen as the working gas, having a simple, well-studied optical spectrum. It is also important that in He processes of step excitation under the conditions studied are substantially attenuated by a significant energy interval between energy levels with the main quantum numbers $n = 2$ and $n \geq 3$ [222].

The values of $\sigma_{k\,max}$ and $v_{e\,max}$ were taken from [223], where the values of these quantities were measured at low pressures $p < 10^{-2}$ Torr. In the range of higher pressures, the values of $\sigma_{k\,max}$ and $v_{e\,max}$ may change somewhat under the influence of secondary processes (stepwise excitation, cascade transitions).

Let us consider the question of the accuracy of the measurements of n_{e0} with the aid of the proposed procedure. We note that, in accordance with formulas (2.15) and (2.19), the technique is not sensitive to the error in setting the value $v_{e\,max}$.

Taking into account that $\sigma_k(v) = \sigma_{k\,max} f_k(v)$, where $f_k(v)$ is the excitation function of the upper level of the spectral line, we obtain according to (2.19):

$$n_{e0} = \frac{I_\lambda^{CL}}{I_\lambda^{nc}} \cdot \frac{\langle \sigma_k(v)v_e \rangle}{\sigma_{k\,max}v_{e0}} \cdot n_e^{nc} = \frac{I_\lambda^{CL}}{I_\lambda^{nc}} \cdot \frac{\langle f_k(v)v_e \rangle}{v_{e0}} \cdot n_e^{nc}. \qquad (2.22)$$

According to (2.22), the method of relative intensities is not sensitive to the error in setting $\sigma_{k\,max}$, in contrast to the method of absolute intensities. Both these methods are sensitive to setting the value of v_{e0}.

Let us dwell on possible errors when using the values of $\sigma_{k\,max}$ and v_{e0} taken from the literature for n_{e0}. Often, the measurements used the He I line 5876 Å. According to [224], when the pressure in the range $p = 10^{-3}$–10^{-1} Torr changes, the measured value of $\sigma_{k\,max}$ for the 5876 Å line changes 2–2.5 times in the direction of increase.

According to the data of [225], the spread of the energies of the emitted electrons for the conditions under consideration is $\varepsilon_{e0} = 4$–12 eV. Assuming $\overline{\varepsilon_{e0}} = 8$ eV in the calculations, we find that the values of $\overline{V_{e0}}$ given here can be maximally different from the real values by 1.2–1.4 times.

The error in the method of measuring n_{e0} from the absolute intensities of the spectral lines by formula (2.15) is determined

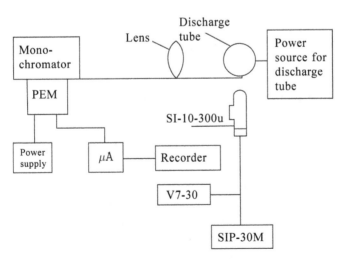

Fig. 2.41. Block diagram of the experimental setup for measuring the concentration of emitted electrons n_{e0}.

mainly by an error in the measurement of the absolute intensities of the lines and is of the order of 100%.

The accuracy of the relative intensity method, in accordance with the expression (2.22), is determined by the accuracy of the probe method in determining n_e and T_e, which according to generally accepted estimates is of the order of 20–30%.

The experimental setup for measuring the concentration of n_{e0} emitted electrons is shown in Fig. 2.41.

The optical path was calibrated using a SI-10-300u light-measuring lamp, the filament of which was installed at a distance from the lens corresponding to the position of the axis of the cylindrical discharge tube. The path was calibrated at points along the wavelengths measured (He *I*, λ = 3889 Å, 4471 Å, 5876 Å, 7065 Å, 7281 Å).

To test the efficiency of the procedure and to measure n_{e0}, a tube with a flat Mo cathode (4.7 cm in diameter), a conical anode of Ni, a cathode and anode cone base of 40 cm was used in the RFCD with internal electrodes. A number of probes were soldered into the tube, at different distances from the cathode (probe diameter 0.1 mm, probe length 5 mm).

To measure the concentration of n_{e0} in the RFCD with external electrodes, a tube of molybdenum glass with glass ends was used.

The He pressure (0.1–1 Torr) used in the experiment was chosen because of the reasonably high signal level in the cathode film, the stability of the positive column (absence of striations), the possibility of using the probe technique for measuring the plasma parameters of DCGD.

The method for determining the brightness $b_\lambda(T)$ of a reference source as a function of the wavelength at a known brightness temperature at the calibration wavelength of a light measuring lamp (λ = 6328 Å) is described in [226]:

1) the true temperature of the tape of lamp *T* is found from the relation:

$$\frac{1}{T} - \frac{1}{T_{br}} = \frac{\lambda}{1.4384} \ln\left[\tau(\lambda)\varepsilon(\lambda,T)\right],$$

where λ is measured in cm; *T* – in K; $\tau(\lambda)$ is the transmittance of the lamp window; $\varepsilon(\lambda,T)$ is the emissivity of tungsten at a given wavelength at a given temperature. The values of $\varepsilon(\lambda,T)$ were used in [227];

2) from the obtained temperature *T*, the brightness of the absolutely black body $b_{ABB}(\lambda,T)$ is found as a function of the wavelength:

$$b_{\text{ABB}}(\lambda,T) = \frac{1.18 \cdot 10^{-5}}{\lambda^5}\left[\exp\left(\frac{1.4384}{\lambda T}-1\right)\right]^{-1};$$

3) the brightness of the reference source was found by the formula:

$$b_\lambda(T) = \tau(\lambda)\varepsilon(\lambda,T)b_{\text{ABB}}(\lambda,T).$$

Writing out the expressions for the radiation flux from the volume source to the input slit of the monochromator and from the reference source [228–230], and taking into account the linearity of the photomultiplier light characteristic, we obtain an expression for the intensity of the spectral line:

$$I_{ki} = \frac{I_{\text{sour}}}{I_{\text{ref}}}\, 8\pi b_\lambda \frac{b'_{\text{ref}}b_{\text{ref}}h_{\text{ref}}}{Db_{\text{sour}}h_{\text{sour}}l_{\text{sour}}}, \qquad (2.23)$$

where I_{sour} and I_{ref} are the currents of the photomultiplier for the investigated radiation source and the reference light source; b_λ – brightness of the reference source at wavelength λ; b'_{ref} is the width of the output slit of the monochromator for the reference source; b_{ref} and h_{ref} are the width and height of the input slit of the monochromator for the reference source; $1/D = d\lambda/dl$ is the inverse linear dispersion of the monochromator; b_{sour} and h_{sour} – the width and height of the entrance slit of the monochromator for the light source under study; l_{sour} – length of the light source.

A lens with a focal length of 30 cm was set between the light source and the input slit of the monochromator at a double focal length so that the source image was projected onto the slit without conversion.

Calculations of the absolute intensities of the spectral lines were made using the formula (2.23), which for the experimental conditions at a wavelength $\lambda = 5876$ Å becomes:

$$I_{ki} = 10^{-7}b_\lambda \frac{I_{\text{sour}}}{I_{\text{ref}}} \text{ (for tube 1),}$$

$$I_{ki} = 1.3\cdot 10^{-7}b_\lambda \frac{I_{\text{sour}}}{I_{\text{ref}}} \text{(for tube 2),}$$

$b_{\text{ref}} = b'_{\text{ref}} = b_{\text{sour}} = 0.1$ mm; $h_{\text{sour}} = h_{\text{ref}} = 2$ mm; $l = 50$ mm for tube 1; $l = 40$ mm for tube 2; $1/D = 20$ Å/mm.

Table 2.1. Values of the concentrations of emitted electrons n_{e0} in DCGD

λ, Å	I_p , mA	ε_{e0}, eV	n_{e0}, cm^{-3}
3889	4	4	$9 \cdot 10^4$
		12	$5 \cdot 10^4$
	8	4	$2 \cdot 10^5$
		12	$1 \cdot 10^5$
4471	4	4	$2 \cdot 10^6$
		12	$9 \cdot 10^5$
	8	4	$6 \cdot 10^6$
		12	$3 \cdot 10^6$
5876	4	4	$1 \cdot 10^6$
		12	$7 \cdot 10^5$
	8	4	$4 \cdot 10^6$
		12	$2 \cdot 10^6$
7065	4	4	$1 \cdot 10^5$
		12	$6 \cdot 10^4$
	8	4	$2 \cdot 10^5$
		12	$8 \cdot 10^4$
7281	4	4	$9 \cdot 10^4$
		12	$5 \cdot 10^4$
	8	4	$2 \cdot 10^5$
		12	$1 \cdot 10^5$

2.11.1. Dependence of the concentration of emitted electrons n_{e0} on RF voltage $V_\sim$ and current $I_\sim$ of the discharge

To test the measurement technique and compare the physical conditions in the RFCD and DCGD, the experiments were performed with both discharges.

To determine n_{e0} in DCGD for two values of the discharge current, the radiation intensity distribution was obtained as a function of the distance to the cathode for spectral lines differing in the course of the excitation function and in the lifetime of the upper excited state τ:

$$\text{He } I \text{ 3889 Å} \quad (2^3S - 3^3P), \quad \text{He } I \text{ 4471 Å} \quad (2^3P - 4^3D),$$
$$\text{He } I \text{ 5876 Å} \quad (2^3P - 3^3D), \quad \text{He } I \text{ 7065 Å} \quad (2^3P - 3^3S),$$
$$\text{He } I \text{ 7281 Å} \quad (2^1P - 3^1S).$$

Table 2.2. Values of the concentrations of emitted electrons n_{e0} in DCGD, measured by three methods

ε_{e0},eV	n_{e0}, cm^{-3}		
	by absolute intensity	by the relative intensity	by measurement of discharge current
4	$8 \cdot 10^4$	$1 \cdot 10^5$	$4 \cdot 10^4$
12	$5 \cdot 10^4$	$8 \cdot 10^4$	$2 \cdot 10^4$

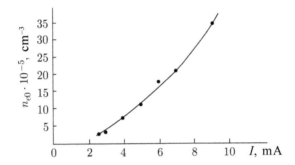

Fig. 2.42. Dependence of the concentration of emitted electrons n_{e0} in DCGD, He, $p = 1$ Torr.

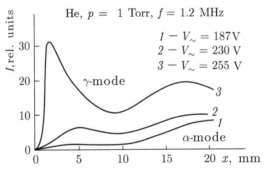

Fig. 2.43. The intensity distribution of the He *I* 5876 Å $I_\lambda(x)$ spectral line in the RFCD with internal electrodes.

The main measurements were carried out on the line He *I* 5876 Å, for which a distinctly pronounced region of CL with an acute maximum was obtained. In addition, the lifetime of the upper excited state of the He *I* 5876 Å line is small [231]. Therefore, the excited atom is illuminated at the same place in the space where it was excited.

The concentrations of the emitted electrons, obtained with the help of measurements of the absolute intensities of the spectral lines for two values of the discharge current $I_p = 4$ mA and 8 mA, are given in Table 2.1.

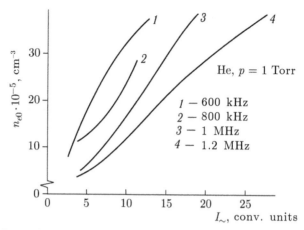

Fig. 2.44. Dependence of the density of emitted electrons n_{e0} in RFCD with internal electrodes on the discharge current.

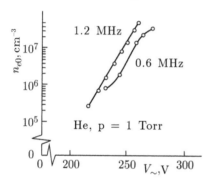

Fig. 2.45. Dependence of the density of emitted electrons n_{e0} on the amplitude of RF voltage $V_\sim$ and the frequency of the field f.

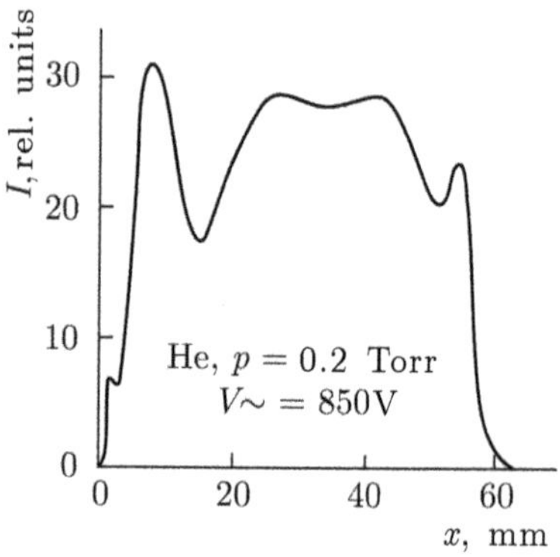

Fig. 2.46. The spatial distribution of the intensity of the spectral line He I 5876 Å $I_\lambda(x)$ in RFCD with external electrodes.

The values of n_{e0} in DCGD for a discharge current of 1 mA, obtained by three methods:

1) from the measured absolute intensity of the spectral line 5876 Å;

2) from the measured relative intensity of the 5876 Å spectral line in the CL region and the positive column and the plasma parameters (n_e, T_e) of the positive column;

3) from measurements of the discharge current to the cathode of DCGD, are given in Table 2.2.

In determining n_{e0} from measurements of the discharge current, the data of Ref. 225 on the values of the mean velocity v_{e0} with which electrons emerge from the cathode and the coefficient of ion–electron

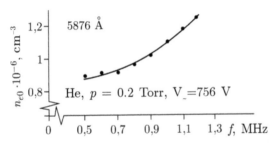

Fig. 2.47. Dependence of the density of emitted electrons n_{e0} in the RFCD on the frequency of the RF field f.

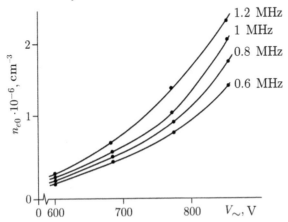

Fig. 2.48. Dependence of the density of emitted electrons n_{e0} in the RFCD with external electrodes on the amplitude of the RF voltage $V_{\sim}$ and the frequency of the RF field f.

emission $\gamma_i = 0.3$ were used. As a result, the values $n_{e0} = (2-4)\cdot10^4$ cm^{-3} were obtained.

Using the measurements of the absolute intensities of the 5876 Å spectral line in DCGD, the dependence of n_{e0} on the magnitude of the discharge current was obtained (Fig. 2.42). At the same time, the discharge current varied within 1–10 mA, since at large discharge currents the sputtering of the cathode significantly hampered the correct optical measurement.

At the RFCD with internal electrodes, the distribution $I_\lambda(x)$ of the emission intensity of the 5876 Å line was obtained for different discharge modes (Fig. 2.43).

In a discharge with internal electrodes, the dependence of n_{e0} on the RF current for different values of the field frequencies (Fig. 2.44) was investigated in terms of the absolute intensity of the spectral line He *I* 5876 Å. The current range is 1–30 mA, the frequency is 0.45–1.2 MHz.

For the RFCD with internal electrodes, the dependence of n_{e0} on the frequency and voltage of the RF field applied to the discharge gap is obtained from the absolute intensities of the 5876 Å spectral line (Fig. 2.45).

The intensity distribution of the 5876 Å spectral line along the discharge gap for RFCD with external electrodes was studied (Fig. 2.46).

The frequency dependence $n_{e0}(f)$ at a fixed amplitude of the RF voltage $V_{\sim} = 765$ V in the frequency range $f = 0.6-1.0$ MHz was investigated (Fig. 2.47).

In the discharge with external electrodes, the dependence $n_{e0}(V_{\sim})$ for different values of the field frequency was obtained (Fig. 2.48).

Let us consider the results of the investigation of electron emission from the cathode of DCGD and the RFCD electrodes.

According to the previously mentioned methodological considerations, the main measurements of the density of the emitted electrodes n_{e0} were made using the spectral line He I 5876 Å.

Comparison of the value of n_{e0} obtained in DCGD using the above three methods reveals their satisfactory agreement (Table 2.2).

Let us analyze the experimental results taking into account the peculiarities of each method used.

The method for determining n_{e0} from the measured value of the discharge current using formula (2.21) requires the use of additional data from the literature – the ion–electron emission coefficient γ_i and the velocities v_{e0} with which the electrons leave the cathode. The drawbacks of this method include its applicability only to DCGD in normal and weakly anomalous modes, the need to draw from different sources not always accurate and accessible data on γ_i and v_{e0}.

The method of relative intensities of determining n_{e0} by formula (2.19) is most simple for the case of DCGD when it is sufficient to measure only the relative intensities of the line I_{ki} in the cathode luminescence region and the positive column, and the values of n_e^{nc} and T_e^{nc} can be calculated from the Schottky theory [151] (in this case, the technique is contactless) or to find n_e^{nc} and T_e^{nc} by the probe method and the rest of the information is taken from the literature.

For the latter method, in the case of the γ-type RFCD, it is necessary to measure n_e^{nc}, T_e^{nc} (in general, $f(\varepsilon)$ is the electron energy distribution function) in the discharge plasma.

Finally, the third method of absolute intensities using formula (2.15) is most preferable for the following reasons. It is contactless and requires measurement of a single quantity – the absolute intensity of the spectral line I_{ki} – and is applicable for measurements in both normal and anomalous DCGD and γ-type RFCD with internal and external electrodes.

The values of n_{e0} obtained by the method of absolute intensities with a methodological purpose (Table 2.1) in DCGD at different physical conditions using a number of spectral lines differ in the same order of magnitude.

The most correct values were n_{e0}, found from the He I line of 5876 Å.

The appreciably underestimated values of n_{e0}, found from the intensity of the He I line of 3889 Å, are explained by the significant reabsorption of this line.

In DCGD it was found that when the current is changed by an order of magnitude, the value of n_{e0} increases, in practice, by the same factor (Fig. 2.42).

A similar dependence of the density of the emitted electrons n_{e0} on the RF current strength was observed in the RFCD with internal electrodes (Fig. 2.44). Although in this case it must be borne in mind that the quantity $I_\sim$, measured by the device, is not equal to the true current flowing through the discharge gap, but also includes parasitic and capacitive currents. Nevertheless, the character of the $n_{e0}(I_\sim)$ dependence is qualitatively reflected correctly in this case.

Let us pay attention to the frequency dependence of the value of n_{e0} for a fixed value of the RF current $I_\sim$. As can be seen in Fig. 2.44, in this case n_{e0} decreases with increasing frequency f.

This can be explained as follows. The total current $I_\sim$ in the near-electrode layer is composed of the conduction and bias currents:

$$I_\sim = I_{cond} + I_{bias}.$$

The concentration of emitted electrons n_{e0} under these conditions is due to the ion flux arriving at the electrode, i.e., the conduction current I_{cond}. As the frequency f grows, the bias current increases

$$I_{bias} = \frac{1}{4\pi} \cdot \frac{dE}{dt} = i\frac{1}{4\pi}\,\omega E \sim fE,$$

so that for $I_\sim = $ const the contribution of I_{bias} to the total current $I_\sim$ increases is proportional to the frequency f.

We note with respect to the component of the conduction current I_{cond} that two opposite tendencies act with increasing frequency: 1) As the frequency f increases, the plasma charge density n_i increases and the current I_{cond} increases; 2) simultaneously with the increase of f, the ion flux to the electrode must decrease due to the increase in the inertia of ions, which leads to a decrease in I_{cond}. The experimental results presented in Fig. 2.44 show, with allowance for what was said above, that at $I_{\sim} = const$, the component of the conduction current I_{cond} decreases with increasing frequency of the RF field, which leads to a decrease in the density of the emitted electrons.

The nature of the dependence of the intensity of emission processes on the internal electrodes of RFCD from the amplitude of the RF voltage $V_{\sim}$ is shown in Fig. 2.45. It can be seen that even in a relatively small range of changes in the parameters of the RFCD, there is a significant change the intensity of electron emission: n_{e0} grows rather rapidly with increasing $V_{\sim}$.

The function $n_{e0}(V_{\sim})$ is found to depend on the frequency of the RF field. Moreover, at $V_{\sim} = const$ an increase in n_{e0} was observed with increasing frequency f, in contrast to the case when $I_{\sim} = const$ (Fig. 2.47). In this case ($V_{\sim} = const$), irrespective of the behaviour of the bias current, the conduction current (and, consequently, the ion flux to the electrode $\sim I_{cond}$) certainly increases with increasing frequency f. This is due to the fact that two factors contribute to the growth of the ion flux to the electrode, as well as to the growth of n_{e0}: 1) the concentration n_i increases with increasing f, 2) in this range of the frequencies with increasing f, the quasi-stationary jumps in the potential U_s in the near-electrode layer increase, which intensifies the ionization processes by electron beams and increases the velocity of the ions to the electrode.

The obtained data for the RFCD with external electrodes are of interest. As can be seen from Figs. 2.47 and 2.48, the dependences $n_{e0}(f)$ and $n_{e0}(V, f)$ have a qualitatively similar character with the corresponding dependences for RFCD with internal electrodes. Absolute values of n_{e0} in the case of RFCD with external electrodes turned out to be somewhat smaller than for RFCD with internal electrodes, which can easily be explained by some differences in the experimental conditions.

According to the measured values, the concentration of the emitted electrons n_{e0} and the plasma electrons in the region of 'negative luminescence' n_e DCGD and RFCD, we estimate the gas gain factors [151] for these discharges

$$K = \frac{n_e}{n_{e0}} = \frac{e^{\alpha d}}{1 - \gamma(e^{\alpha d} - 1)},$$

where d is the distance from the electrode to the region of 'negative glow.'

According to the experimental data [232]: DCGD in He, p = 0.58 Torr, I_d = 0.6 mA, cathode – Mo, cathode diameter 2.7 cm, n_e = 6 · 10^9 cm^{-3}, we will accept the value of the velocity of the emitted electrons ε_{e0} = 9 eV [225], $\gamma \sim 10^{-1}$, then we get the value

$$n_{e0} = \frac{I_d \gamma}{S_{cat} e v_{e0}} = 1.7 \cdot 10^5 \text{ cm}^{-3}.$$

From this

$$K = \frac{n_e}{n_{e0}} = \frac{6 \cdot 10^9}{1.7 \cdot 10^5} = 1.8 \cdot 10^4.$$

According to our experimental data:

1) DCGD, He, p = 1 Torr, I_d = 1 mA, n_e = 9 · 10^8 cm^{-3}, cathode diameter 4.7 cm, Mo cathode, n_{e0} = 6 · 10^4 cm^{-3}. From this

$$K = \frac{9 \cdot 10^8}{6 \cdot 10^4} = 1.5 \cdot 10^4.$$

2) RFCD, He, p = 1 Torr, $V_{\sim}$ = 220 V, f = 1.2 MHz, cathode diameter 4.7 cm, Mo electrode,

$$n_e = 1.5 \cdot 10^9 \text{ cm}^{-3}, \quad n_{e0} = 4 \cdot 10^5 \text{ cm}^{-3}.$$

From this

$$K = \frac{1.5 \cdot 10^9}{4 \cdot 10^5} = 0.4 \cdot 10^4.$$

Thus, the intensity of the processes of avalanche ionization in the near-electrode regions of RFCD and DCGD under close physical conditions is of the same order of magnitude.

According to what was said earlier, the spectroscopic measurement method of the density of the emitted electrons n_{e0} has significant advantages over the mentioned electric method, since: 1) it is not necessary to know the problem value of the surface ionization coefficient γ, 2) the electrode material (metal, dielectric, semiconductor) does not matter, 3) to measure the emission current density $j_{e0} = en_{e0} v_{e0}$ only the absolute intensity of the spectral line I_λ and purely spectroscopic reference data are needed, 4) the method is applicable to RFCD with both internal and external electrodes.

With the help of this spectroscopic method, in general, one can determine the Townsend coefficient γ using the materials in this monograph.

The importance of the spectroscopic method is also increased due to the fact that in low-pressure RFCD the density of emitted electrons is equal to the density of the near-electrode electron beam $n_{e0} = n_{eb}$.

The spectroscopic method of measuring the value of n_{e0}, depending on all the discharge parameters, makes it possible to optimize the electron-emission processes at the electrodes and the RFCD modes as a whole.

2.12. Contactless method for diagnostics of RFCD NESCL parameters

Experimental studies of the physical characteristics of the boundary layers of space charge are important for studying the mechanism of RF capacitive discharge (RFCD), widely used in practice, and for organizing the effective control of plasma chemical processes in the production of microelectronics products. In industrial systems with the arrangement of processed substrates on the discharge electrodes, the near-electrode charge-space layers (NESCL) are directly adjacent to the surface of the processed plates, therefore the study of the characteristics of the near-electrode region provides the necessary monitoring of the internal parameters of the system characterizing the state of the working medium directly in the processing zone. Obviously, only such a control can ensure a high level of reproducibility of the technology.

In [206], the method of contactless control of the quasi-stationary characteristics of the NESCL of the RFCD was proposed: a quasi-stationary jump of the potential U_s on the NESCL and the thickness d_s of this layer. In the case of a collisionless near-electrode layer, often encountered in practice, the method makes it possible to determine also the current density of positive ions j_{is} from the plasma to the electrode and the concentration of charges $n_{es} = n_{is}$ on boundary of the NESCL.

The proposed method assumes averaging of the investigated quantities in time and space. Therefore, the smallest measurement

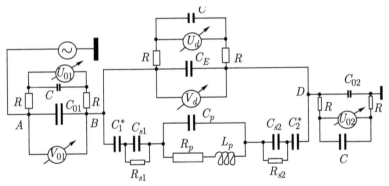

Fig. 2.49. Equivalent circuit of the experimental electric circuit of RFCD. U_{01}, U_{02}, U_d – voltmeters for measuring quasi-stationary voltages; V_{01}, V_d – voltmeters for measuring alternating voltages; R_p, L_p, C_p – active resistance, inductance and plasma capacitance of RFCD, respectively.

errors occur in the diagnosis of reactors with a high surface uniformity of plate processing.

The method was experimentally tested under conditions of RFCD in helium at a gas pressure of 0.3–0.5 mm Hg in the frequency range of the RF field $f = 0.64$–2 MHz. The experiments were carried out in a discharge tube with internal electrodes with simultaneous measurements using the Langmuir probe technique. Reasonably good correspondence of the results obtained by the two methods (within 10–15% for the values U_s) gave reason to use the contactless method for studying RFCD with external electrodes, where previously no experimental methods of such measurements were available.

Consider a RFCD with flat electrodes E_1 and E_2 of area S_1 and S_2, covered with a layer of a dielectric that creates the capacitances C_1^* and C_2^* between the electrodes and the imaginary second plates from the side of the interelectrode space (Fig. 2.23). The RF generator and the electrode E_1 are separated by the capacitor C_{01}. The second electrode E_2 is grounded through the separating capacitor C_{02}, which is introduced to implement the proposed measurement method. The physical scheme of the RFCD is shown in Fig. 2.23. Here, the plasma is shown in the form of a shaded region of the discharge gap separated from the electrodes by layers of space charge. Their dimensions d_{s1} and d_{s2} represent the quasi-stationary thicknesses of the NESCL.

The schematic diagram of the electric circuit of the RFCD, used in this method, is shown in Fig. 2.49.

The approach used in this method is based on three basic assumptions:

1) the quasi-stationary nature of the ionic conduction current in the layer

$$1/\tau < \omega \ll \omega_{Le},\qquad(2.24)$$

where τ is the time of ion motion in the near-electrode layers, $\omega = 2\pi f$ is the angular frequency of the RF field, ω_{Le} is the Langmuir frequency of the plasma electrons at the boundary of the layer;

2) the prevailing role of the bias current in the layer due to the ratio of the active and capacitive resistances of the NESCL:

$$R_{s1,2} \gg \frac{1}{\omega C_{s1,2}},\qquad(2.25)$$

which makes it possible to represent the NESCL in the equivalent electric discharge circuit in the form of capacitances $C_{s1,2}$ (Figs. 2.23, 2.49);

3) the negligibly small contribution of charged particles introduced into the dielectric permittivity ε_s in the layer, i.e., $\varepsilon_s \approx 1$.

Then we can formulate the following system of equations:

$$q_1 = U_{s1}C_{s1} = U_1^*C_1^* = U_{01}C_{01},\qquad(2.26)$$

$$q_2 = U_{s2}C_{s2} = U_2^*C_2^* = U_{02}C_{02},\qquad(2.27)$$

$$U_d + U_{s1} - U_{s2} - U_1^* + U_2^* = 0,\qquad(2.28)$$

$$\frac{V_{01}}{V_d} = \frac{C_d}{C_{01}},\qquad(2.29)$$

where q_1 and q_2 are the quasi-stationary values of the charges arising in the capacitances C_{s1} and C_{s2} and electrostatically induced according to the law of induction on adjacent capacitors C_1^*, C_2^* and C_{01}, C_{02}, respectively; U_{s1}, U_{s2} are quasi-stationary potential jumps in the corresponding NESCL; U_d is the total drop in the quasi-stationary voltage at discharge; V_{01} and V_d – drops of alternating voltage on the capacitances of the capacitor C_{01} and discharge C_d (between points B and D in Fig. 2.49) respectively.

Equation (2.28) represents the balance of the electrostatic voltage in the discharge with the interelectrode distance d, and relation (2.29) expresses the distribution of the alternating voltage in the circuit segment between points B and D and capacitor C_{01} (Fig. 2.49). At the same time, we have a capacitive divider of the RF voltage falling in the discharge circuit.

The unknown values in the system of equations (2.26)–(2.29) are the values of the parameters of the near-electrode discharge layers U_{s1}, U_{s2}, C_{s1} and C_{s2}.

Combining the expressions (2.26)–(2.29), we obtain:

$$U_{s2} = U_{s1} + U_d - \frac{U_{01}C_{01}}{C_1^*} + \frac{U_{02}C_{02}}{C_2^*}. \qquad (2.30)$$

The total capacitance of the discharge C_d can be represented by the expression

$$C_d = (C_{s1}^{-1} + C_{s2}^{-1} + C_1^{*-1} + C_2^{*-1})^{-1} + C_E,$$

where C_E is the interelectrode capacitance parallel to the discharge. With considering of the last formula, using (2.26), (2.27), (2.29), and (2.30), we obtain

$$U_{s1} = [\frac{V_d}{V_0 C_{01} - V_d C_E} + C_1^{*-1}(\frac{U_{01}C_{01}}{U_{02}C_{02}} - 1) -$$
$$\frac{U_d}{U_{02}C_{02}} - \frac{2}{C_2^*}] \times (\frac{1}{U_{01}C_{01}} + \frac{1}{U_{02}C_{02}})^{-1}. \qquad (2.31)$$

We find the thicknesses of the near-electrode layers d_{s1} and d_{s2}, using (2.26) and (2.27) and the formula for the capacitance of a flat capacitor $C_s = \varepsilon_s S/4\pi d_s$.

As a result, taking into account the third initial assumption, we get:

$$d_{s1} = S_1 U_{s1} / 4\pi U_{01} C_{01}, \qquad (2.32)$$

$$d_{s2} = S_2 U_{s2} / 4\pi U_{02} C_{02}. \qquad (2.33)$$

In the expressions (2.28)–(2.31), the measured quantities are: constant and alternating voltage on the capacitor C_{01} and on the discharge electrodes, as well as the DC voltage across the capacitor C_{02}.

Passing from the measurements in this method using the physical model of the electric circuit of RFCD to the real experimental scheme and measurement procedure, the validity of the proposed method will be briefly justified.

The physical idea of the representation of a NESCL in the form of capacitances in the case of the fulfillment of the condition (2.25) was first proposed in [233], and then experimentally confirmed by the radio engineering method of the 'ionization condenser' [234].

It is also assumed here that the active resistance of the plasma column R_p is significantly lower than the reactive resistances of the plasma and the resistances of the NESCL:

$$R_p \ll \frac{1}{\omega C_p}, \quad R_p \ll \omega L_p \text{ and } R_p \ll R_{s1,2}, \quad R_p \ll \frac{1}{\omega C_{s1,2}}, \quad (2.34)$$

where C_p and L_p are the capacitance and inductance of plasma, respectively.

In order to justify the adopted model representations, we make a number of quantitative estimates for the experimental conditions studied: He, $p = 0.5$ Torr, $f = 2$ MHz, $\omega = 1.3 \cdot 10^7$ s^{-1}, $d_{s1.2} \approx 1$ cm; field strength in the layer $E_s \sim 100$ V/cm, the collision frequency of electrons with atoms $v_{en} \approx 10^9$ s^{-1} and ions with atoms $v_{in} \approx 10^7$ s^{-1}, in the used range of RF voltage, the charge density of the plasma at the centre of the discharge is $n_e \approx 10^8 - 10^9$ cm^{-3} [4], the mass of the ion $M = 6.64 \cdot 10^{-24}$ g.

First, let us verify the fulfillment of inequality (2.24): $\tau = d_s / v_i$, where the mean ion velocity in the layer $v_i = b_i E_s$; the mobility of the ion $b_i = e/Mv_{in}$. Hence: $\tau \approx 5 \cdot 10^{-7}$ s and $1/\tau = 2 \cdot 10^6$ s$^{-1} < \omega$.

According to the above densities of plasma charges n_e, $0.6 \cdot 10^9$ s$^{-1} \le \omega_{Le} \le 1.9 \cdot 10^9$ s^{-1} and, consequently, $\omega_{Le} \gg \omega$.

Thus, inequality (2.24) is satisfied.

Let us now verify the fulfillment of inequality (2.25), taking into account the known ion density in the NESCL $n_{is} \approx 10^7 - 10^8$ cm^{-3} and the electrode area $S = 78$ cm^2.

Taking into account the value of the capacitance of the near-electrode layer $C_s = S/4\pi d_s = 7$ pF, we obtain the value of the capacitive resistance of the NESCL:

$$1/\omega C_s \approx 10 \text{ kOhm.}$$

Let us estimate the active resistance of the NESCL: $R_s = d_s/\sigma_{is} S$, where $\sigma_{is} = e^2 n_{is}/Mv_{in}$ is the specific active conductivity of the layer. Hence we obtain: $R_s \gg 30$ kOhm, which confirms the fulfillment of inequality (2.25).

According to [204], the main contribution to the permittivity in the layer is given by ions:

$$\varepsilon_s = 1 - \frac{\omega_{Li}^2}{\omega^2 + v_{in}^2},$$

where ω_{Li} is the ion Langmuir frequency. Moreover, for the conditions of the carried out verification of the proposed method, it follows from the experimental data of [67] that

$$\omega_{Li}^2 \ll \omega^2 + v_{in}^2 \quad \text{and} \quad \varepsilon_s \approx 1.$$

Let us show that the inequalities (2.34) are satisfied under the conditions under consideration.

Let us compare the active R_p and reactive $1/\omega C_p$ and ωL_p plasma resistances.

First we consider the expression for the plasma capacity

$$C_p = \varepsilon_p S / 4\pi d_p,$$

where $\varepsilon_p \approx 1 - \omega_{Le}^2/(\omega^2 + v_{en}^2)$ is the dielectric constant of the plasma [204], d_p and S are the length and cross section of the plasma column, respectively.

Note that when

$$\omega_{Le} \le v_{en}, \quad \text{then} \quad \varepsilon_p > 0,$$

and the reactance of the plasma has a capacitive character.

In the case of inequality

$$\omega_{Le}^2 > \omega^2 + v_{en}^2, \quad \varepsilon_p < 0,$$

and the nature of reactive plasma resistance is inductive.

Taking into account the parameters of the discharge under consideration, it is easy to estimate the ratio of active and capacitive resistances plasma

$$\frac{R_p}{1/\omega C_p} \le 2 \cdot 10^{-2},$$

as well as active and inductive resistances of plasma

$$2 \cdot 10^{-3} \le \frac{R_p}{1/\omega L_p} \le 9 \cdot 10^{-3}.$$

Similarly, estimates of the ratios of the active resistance of the plasma to the capacitive and active resistances of the NESCL lead to the following results:

$$2 \cdot 10^{-1} \leq \frac{R_p}{1/\omega C_s} \leq 2 \cdot 10^{-2}$$

and

$$\frac{R_p}{R_s} \approx 10^{-2}.$$

Thus, under the experimental conditions, the assumed model relations (2.34) are also satisfied.

NESCLs form in the process of electric breakdown of the discharge gap and the formation of the discharge and ensure the output of RFCD to the stationary mode. As was experimentally established [235], in the steady-state RFCD for each period of the RF field ($T \leq 10^{-6}$ s), the total electric charge arriving at the electrode is zero. Correspondingly, the quasi-stationary parameters of the NESCL capacitor: capacitance C_s, charge q_s and layer thickness d_s remain constant. In this case, the charge q_s provides a quasi-stationary potential difference U_s in the NESCL, due to which during the RF field the charge of the electrons arriving from the plasma to the electrode compensates for the charge of positive ions and the charge of electrons emitted from the surface of the electrode.

In the section of the electric circuit of the RFCD (Fig. 2.49), consisting of the series-connected capacitors C_{01}, C_1^* and C_{s1}, the active element is the capacitance C_{s1}, charge on the plates of which q_{s1} determines the physical processes in the NESCL.

In this case, this quasi-stationary charge q_{s1} is set in all series-connected capacitances of the electrical circuit, including on the measuring capacitance C_{01}.

To measure the quasi-stationary voltages U_{01}, U_{02} and U_d the experimental scheme included voltmeters of an electrostatic system of the S-95 type with a small input capacitance $C' = 5$ pF. The measuring circuit of these devices is open with a direct current, so that there is no leakage of the charge of the capacitance under study through the measurement circuit. In order to exclude the influence of RF voltage on the voltmeter readings, an integrating electrical circuit was used, consisting of two active resistances R and a capacitance C with specially selected values (Fig. 2.49).

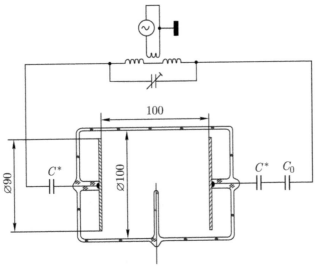

Fig. 2.50. Experimental scheme of a discharge tube with internal electrodes for conducting measurements by contactless and probe methods.

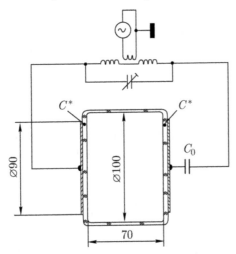

Fig. 2.51. Experimental scheme of a discharge tube with external electrodes for performing measurements of the parameters of the NESCL by the contactless method.

In order to avoid the possible influence of the DC electric circuits of the output stage of the RF generator on the operation of the experimental system, the RF power supply to the discharge gap was provided by a transformer method with no induction (Figs. 2.50, 2.51).

High-frequency voltages V_{01} and V_d were measured by means of electronic voltmeters of type V7-15 with high internal resistance, equipped with a remote high-resistance head with a low input capacitance and a voltage divider DN-2, which made it possible to operate over a wide frequency range (5 kHz–300 MHz) and voltages ($V \leq 1$ kV). Similar measurements were also made with S-95 voltmeters.

Thus, the proposed method includes measuring the electrical voltage on several external elements of the RFCD circuit and calculating the desired parameters by the above formulas using known design parameters of the experimental system.

It is easy to see that in the special cases of the discharge geometry ($C_E \approx 0$) or the absence of a dielectric on the surface of the electrodes or a thin layer of it, when the values of C_1^* and C_2^* are large, expressions (2.26)–(2.29) are substantially simplified.

If in the near-electrode discharge layers the positively charged ions move without collisions with atoms, which occurs when the mean free path of ions is $\lambda_i = 1/n_a Q_{ia} \geq d_s$, where n_a is the concentration of atoms, and Q_{ia} is the cross section for resonant charge exchange of ions (usually this condition is satisfied at a gas pressure $p \leq 10^{-1}$ mm Hg), then the current density of the ions on the electrodes can be estimated using the Child–Langmuir expression [4]

$$j_i = \frac{1}{9\pi}\sqrt{\frac{2e}{M_i}}\frac{U_s^{3/2}}{d_s^2},$$

where e is the electron charge, and M_i is the ion mass. At a known electron temperature T_e, according to Bohm's expression [4]

$$j_i = 0.8\,en_i\sqrt{\frac{2kT_e}{M_i}}$$

it is possible to estimate the concentration of ions at the boundary of a quasi-neutral plasma:

$$n_i = 1.25\,\frac{j_i}{e}\sqrt{\frac{2kT_e}{M_i}}.$$

The method was verified out in a symmetrical RFCD with flat internal and external electrodes (Figs. 2.50 and 2.51).

In the experiments carried out, helium-filled sealed gas-discharge tubes having internal planar electrodes 90 mm in diameter with an interelectrode distance of 100 mm and external electrodes 90 mm in diameter with an interelectrode distance of 78 mm, separated from the discharge space by optically transparent quartz 4 mm thick, were used. In both types of discharge the condition

$$C_E \ll C_0 V_0 / V_d$$

was satisfied.

In this case, the expressions (2.30) and (2.31) are reduced for discharge with external electrodes to a single formula $U_s = U_0 C_0 (V_d/2V_0 C_0 - 1/C^*)$, which is further simplified in the case of a discharge with internal electrodes: $U_s = U_0 V_d / 2V_0$, and the thickness of the NESCL for both types of discharge is determined by the formula

$$d_s = S U_s / 4\pi U_0 C_0.$$

The first stage of the experiments was performed in the RFCD with internal electrodes at a gas pressure of 0.3–0.5 mm Hg at field frequencies in the range of 0.64–2 MHz. Here, measurements of quasi-stationary voltages with the contactless method were monitored using an independent technique – a special probe circuit consisting of a micro-ammeter of the M-193 type shunted by capacitance and high resistance resistors, the selection of parameters of which ensured increased accuracy of measurements. In this case, the Langmuir probe was placed in the centre of the discharge gap.

In order to verify the absence of influence of instrument effects on the results of measurements by a non-contact method, the dependence of the resulting quasi-stationary voltage U_s on the value of the measuring capacitance C_{01} is experimentally verified. During operation, the C_{01} values varied within 10–30 000 pF. The discrepancies in the U_s values measured by the two methods were 10–15%, regardless of the capacitance C_{01}.

We estimate and compare the duration of the RF field $T \approx 10^{-6}$ s with the characteristic discharge times τ_{01} and τ_{02} of the measuring capacitances $C_{01} = C_{02} = 50$ pF, taking into account the equivalent experimental electrical circuit (Fig. 2.49), where the voltmeters of the electrostatic system U_{01}, U_{02} and U_d break the DC circuit, and the electronic voltmeters V_{01} and V_d have internal resistances $R_{01} = R_d = R_V \approx 10^6$ Ohm.

a) Capacity C_{01} (RFCD with external and internal electrodes)

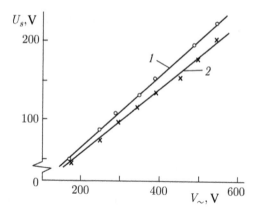

Fig. 2.52. Dependence of the jump in the quasi-stationary voltage U in the NESCL of the discharge with internal electrodes on the amplitude of the applied RF voltage $V_\sim$. RFCD in helium, $p = 0.5$ mm Hg. $f = 1$ MHz, $C = 10$ pF. *1* – probe method, *2* – contactless method.

$$\tau_{01} = R_{01}C_{01} = R_V C_{01} \approx 10^6 \cdot 50 \cdot 10^{-12} = 5 \cdot 10^{-5} \text{ s, i.e. } \tau_{01} \gg T.$$

b) Capacity C_{02}
    – RFCD with external electrodes

$$\tau_{02} \approx (R_d + R_{01})C_{02} = 2R_V C_{02} \approx 2 \cdot 10^6 \cdot 50 \cdot 10^{-12} = 10^{-4} \text{ s} \gg T.$$

    – RFCD with internal electrodes

$$\tau_{02} \approx (2R_s + R_p + R_{01})C_{02} \approx R_V C_{02} = 10^6 \cdot 50 \cdot 10^{-12} = 5 \cdot 10^{-5} \text{ s} \gg T.$$

Thus, during the period of the RF field the charges of capacitances C_{01} and C_{02} practically do not change, and the measured quasi-stationary voltages U_{01} and U_{02} are close to constant.

The data presented indicate the correctness of the procedure for measuring the quasi-stationary potential difference in the NESCL in the proposed contactless method.

The results of measurements by both methods are shown in Fig. 2.52. Their agreement can be considered satisfactory, taking into account the following systematic measurement errors and the fact that the probe measurements are local in the centre of the discharge on the axis of the tube.

Before conducting the measurements in the RFCD with external electrodes, a model experiment was conducted. In this case, the

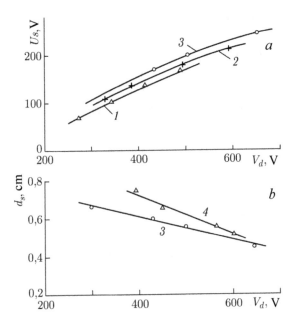

Fig. 2.53. Dependence of the parameters of the NESCL of a discharge with external electrodes from the amplitude V and the frequency f of the applied RF voltage RFCD in helium, $p = 0.\ 5$ mm Hg, $C = 10$ pF. The frequency of the field is f: $1 - 0.64$ MHz, $2 - 1.0$ MHz, $3 - 2.0$ MHz, $4 - 1.6$ MHz. a) The dependence of the near-electrode quasi-stationary jump of the potential U on V; b) the dependence of the thickness of the NESCL d on V.

RFCD with internal electrodes was used, and the dielectric coating of the external electrodes was modeled by capacitors whose capacitance ($C^* = 50$ pF) was equal to the measured capacity of the end face of the quartz wall of the discharge tube (Fig. 2.50). The values of U_s measured in this experiment by the probe and contactless method differed approximately by 10%.

In a discharge with external electrodes, of course, the parameters of the NESCL were determined only by the contactless method. The results of these measurements are shown in Fig. 2.53, a, b, whence it is seen that the obtained values of the parameters of the NESCL U_s and d_s are close to the corresponding values in the RFCD with internal electrodes, the value U_s in the same way as it was established earlier [67], increases with the frequency of the RF field in the given band.

Strictly speaking, a reliable comparison of the values of the jump in the quasi-stationary potential U_s on the boundary layers of the discharge, obtained by probe and contactless methods, is strictly possible only with the known distribution of the electrostatic potential in the discharge. Since no such studies have been carried out in the present study, relevant data from the literature were used for comparative analysis. In this sense, the experiment described in Ref. [51] is of interest, whence it follows that there may be some difference between the quasi-stationary electric potentials between the discharge center and the NESCL boundary. In the proposed contactless method this was not taken into account.

As can be seen in Fig. 2.52, the discrepancy between the data of the two methods used is systematic: the results of the contactless method are constantly underestimated with respect to the data of the probe method. In addition to the above, there is at least one more reason explaining the observed trend. The fact is that in the electric equivalent circuit of the RFCD the near-electrode layer is represented as a flat capacitor. Meanwhile, in real conditions flat is only one of its plates – the surface of the electrode. The second plate is substantially curved in connection with the inhomogeneous radial distribution of the density of charged particles in the discharge tube. Obviously, the capacitance of a real capacitor will be lower, and the RF voltage drop will be higher, and the potential drop U_s, caused by the phenomenon of 'RF detection' [28], will also increase. Thus, correction of the parameters of the near-electrode capacitor C_s should reduce the observed divergence by 10–15% with the proven probe method.

In conclusion, it can be stated that the proposed non-perturbing contactless method proved to be efficient in the process of experimental verification and made it possible to measure for the first time the parameters of the near-electrode region of RFCD with external electrodes, which are of considerable interest, in particular, for plasma-chemical technology.

Edge effects in the
RF capacitive discharge

3.1. The near-electrode space charge layer as a source of the edge effects of the RF capacitive discharge

The existence of the near-electrode space charge layer (NESCL) leads to the appearance of a number of edge effects of RFCD.

The electric field present in the RFCD is concentrated in the near-electrode regions of the α-discharge, where the maximum ionization frequency v_i is observed, and the RFCD of the γ-type is characterized by the formation of pulses of the quasi-stationary field in the NESCL, leading to an increase in ionization in the near-electrode regions and intensification of the escape of the plasma charges to the electrodes. In the latter, because of the appearance of significant gradients of the RF field strength, 'anomalous' diffusion of charges [236] and a ponderomotive unidirectional force [25], which influences the motion of charges in near-electrode regions, can also arise.

Owing to the presence of electrons emitted from the electrode surface and significant accelerating quasi-stationary voltages in the NESCL, the appearance of near-electrode high-energy electron beams occurs in the γ-type RFCD [47, 50]. When certain parameters of these beams reach the specified physical conditions, a beam-plasma instability (BPI) can arise in the discharge, a powerful consequence of the appearance of edge beams [145, 169]. Due to the global role of the phenomenon of the appearance of electron beams in the mechanism of low-pressure RFCD and its universal nature, the following chapter will be devoted entirely to this issue.

The edge effect affecting the parameters of the near-electrode plasma is the stochastic heating of the plasma electrons by the oscillating outer boundary of the NESCL [56]. It is also possible to cool a certain group of plasma electrons [6].

A very peculiar edge effect in RFCD is the formation of the spatially extended and structured 'NESCL–plasma' boundary, the dynamic regime of which can lead to a new physical phenomenon – the discrete mechanism of transfer of plasma electrons to the electrodes [145, 237].

The dynamics of the parameters of the NESCL leads to a corresponding time course of physical processes in the discharge gap.

The experimentally recorded edge effect of the RF field in the vicinity of the electrodes also belongs to the number of edge effects of RFCD, as a result of which secondary radial beams emerge from the walls of the discharge tube in the near-electrode region [189].

Under appropriate conditions, the NESCL and the near-electrode plasma in the RFCD can form a sequential oscillatory circuit, as a result of which an intense generation of the electromagnetic field occurs at frequencies substantially exceeding the RF voltage frequency maintaining the discharge [188].

Finally, under the conditions of the γ-type RFCD, a specific character of the weakening of the RF pulses of a quasi-stationary voltage by the near-electrode plasma can be seen [145].

Next, let us consider the edge effects of RFCD in more detail.

3.2. Influence of the NESCL on the transport of charged particles in the near-electrode plasma

The study of charge transport processes in near-electrode regions is an important aspect of the investigation of the physical mechanism of RFCD. This problem is of great importance for the development of a variety of technological applications.

The transport of charges in the NESCL has been little studied both in the collisionless layer and in the diffusion regime.

Up to the present time, a traditional model of plasma oscillating between the electrodes has been used [6].

The processes of electron emission from the surface of electrodes of the low- and medium-pressure RFCD have not been investigated experimentally as yet.

The needs of the development of modern technologies stimulated the study, first of all, of the behaviour of ions in the NESCL ions.

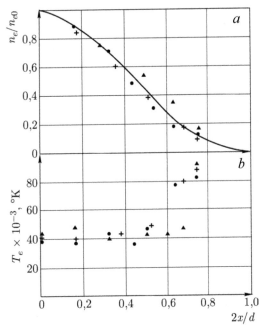

Fig. 3.1. Distribution of plasma parameters of RFCD along the discharge gap at low concentrations of charged particles ($n(0) = 10^8$ cm^{-3}). *a*) distribution $n_e/n(0)$ ($2x / d$); the experimental points correspond to the following conditions: + ($\omega = 1.5$ MHz, $V_\sim = 280$ V, n_e (0) = $1.3 \cdot 10^8$ cm^{-3}); ●($\omega = 3$ MHz, $V_\sim = 220$ V, $n_e(0) = 1.5 \cdot 10^8$ cm^{-3}); ▲ ($\omega = 5$ MHz, $V_\sim = 200$ V, $n_e(0) = 4 \cdot 10^8$ cm^{-3}); the solid line is the curve of 'transition' diffusion [239]. *b*) distribution of T_e ($2x / d$); the experimental values of T_e correspond to the conditions specified in point *a* in the RFCD.

At the same time, the energy spectrum of ions arriving at the surface being treated was in the centre of attention [208, 210, 238].

Since the character of the ion motion in the NESCL of the RFCD is better known, the main attention will be paid to the study of electron transport processes in the near-electrode layers.

For the purpose of experimental study of the effect of boundary conditions on processes in the volume of the discharge, spatial distributions of plasma parameters for various voltages and frequencies of the RF field are investigated in the RFCD in He at pressure $p = 0.5$ Torr (Figs. 3.1–3.5).

The results of the study of the $n_e(x)$ concentration distributions and the electron temperature $T_e(x)$ over the length of the discharge gap at reduced charge densities of n_e plasma are shown in Fig. 3.1. Here, too, the theoretical distribution curve $n_e(x)$ is given for the case of 'transition' diffusion [239], when in the centre of the discharge the electron diffusion is close to the ambipolar one, and in the near-electrode regions to the free one.

As can be seen from the graphs given, for small values of $V_\sim$ and electron concentration (at the centre of the discharge, $n_{e0} = 1.3 \times 10^8$ cm^{-3}), the $n_e(x)$ distribution is close to the 'transition' diffusion curve, and T_e practically does not change in space, except for the vicinity of the electrodes.

The 'transition' diffusion curve was obtained by computer calculation under the assumption that no field is acting in the investigated direction except for the space-charge field of the plasma. Such conditions occur, for example, in the plasma of a positive column of a glow discharge in the radial direction.

In the investigated RF discharge, it was possible to obtain a plasma with a given concentration n_e sufficient to establish the regime of ambipolar diffusion ($n_e \geq 10^9$ cm^{-3}) at different amplitudes of the RF voltage.

Let us consider separately the cases of small ($V_\sim = 100$ V) and higher ($V_\sim \geq 300$ V) RF voltages.

The distributions $n_e(x)$, $T_e(x)$ corresponding to the first case are shown in Fig. 3.2. For comparison, the $n_e(x)$ distribution, obtained theoretically for the case of ambipolar diffusion, is assumed at the same time as the experimental one under the assumption that T_e is constant throughout the discharge gap of plane geometry. The theoretical distribution in this case has the form [157]:

$$n_e(x) = n_{e0} \cos (x / \Lambda), \tag{3.1}$$

where n_{e0} is the concentration at the centre of the discharge, $\Lambda = d/\pi$ is the characteristic diffusion length of the discharge gap, d is the distance between the electrodes.

As can be seen from Fig. 3.2, the experimental distribution of $n_e(x)$ is close to the theoretical one. The electron temperature in this case changed little in space.

Thus, the nature of the motion of charged particles in the longitudinal direction in RFCD at small $V_\sim$, both at low n_e and at concentrations sufficient to establish the mode of ambipolar diffusion, is analogous to the diffusion motion of charges in a plasma between plane electrodes unperturbed by edge effects.

The use of elevated RF voltages leads to a change in the spatial distribution of the plasma parameters. In this case, as established, the electron energy distribution function $f_e(\varepsilon)$ becomes different from the Maxwellian, being enriched by 'fast' electrons. In this connection, the effective electron temperature $T_{e\,\mathrm{eff}}$, determined using the experimentally measured function $f_e(\varepsilon)$, was introduced:

$$T_{e\,\mathrm{eff}} = \frac{2}{3}\frac{\langle \varepsilon_e \rangle}{k},$$

where $\langle \varepsilon_e \rangle = \dfrac{\displaystyle\int_0^\infty \varepsilon_e f_e(\varepsilon)d\varepsilon}{\displaystyle\int_0^\infty f_e(\varepsilon)d\varepsilon}$, k is the Boltzmann constant.

Information on the electronic energy spectrum (EES) of the plasma of the RFCD studied is discussed in detail below. The obtained experimental spatial distributions of n_e and $T_{e\,\mathrm{eff}}$ are shown in Fig. 3.3.

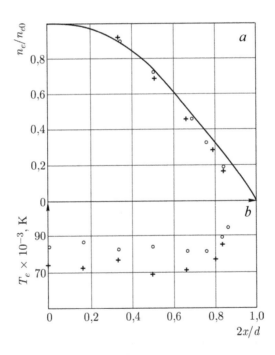

Fig. 3.2. Distribution of plasma parameters of the RF discharge along the discharge gap (ambipolar diffusion mode, relatively small values $V_\sim$). a) distribution $n_e/n_e(0)$ $(2x/d)$; the experimental points correspond to the following conditions: o ($\omega = 12$ MHz, $V_\sim = 160$ V, $n_e(0) = 1.8 \cdot 10^9$ cm^{-3}); + ($\omega = 10$ MHz, $V_\sim = 200$ V, $n_e(0) = 2 \cdot 10^9$ cm^{-3}); the solid line is the theoretical distribution $n_e/n_e(0)$ $(2x/d) = \cos(x/\Lambda)$. b) distribution of $T_e(2x/d)$; the experimental values of T_e correspond to the conditions specified in point *a* in the RF discharge.

From this it is seen that at $V_\sim > 300$ V, the distribution $n_e(x)$ acquires a specific 'bell-shaped' character, differing substantially from the cosine distribution at reduced $V_\sim$.

We note that in the latter case the distribution $T_{e\ eff}(x)$ becomes essentially non-uniform in the discharge gap.

For completeness, let us consider still the experimental results obtained in the study of the radial distributions of the plasma parameters and the corresponding distributions in an asymmetric discharge.

Studies of radial distributions of concentration n_e and T_e temperatures were made for a sufficiently dense plasma. In this case, it is possible to compare the experimental distribution $n_e(r)$ with the theoretical curve of the radial distribution in the positive column of a glow discharge – the Schottky distribution [1]:

$$n_e(r) = n_{e0} J_0\left(2,405\frac{r}{R}\right),$$

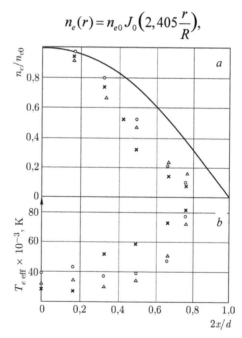

Fig. 3.3. Distribution of the plasma parameters of RF discharge along the discharge gap (the regime of ambipolar diffusion, increased values of $V_\sim$). a) distribution $n_e/n_e(0)$ $(2x/d)$; the experimental points correspond the following conditions: ○ (ω = 10 MHz, $V_\sim$ = 400 V, n_e (0) = 1.1 · 10^{10} cm^{-3}); Δ (ω = 3 MHz, $V_\sim$ = 440 V, $n_e(0)$ = 1 · 10^9 cm^{-3}); × (ω = 1.5 MHz, $V_\sim$ = 680 V, $n_e(0)$ = 1.9 · 10^9 cm^{-3}); the solid line is the theoretical distribution n_e/n_e (0) $(2x/d)$ = cos (x/Λ). b) distribution of $T_{e\ eff}(2x/d)$; the experimental values of $T_{e\ eff}$ correspond to the conditions specified in point *a* in the RF discharge.

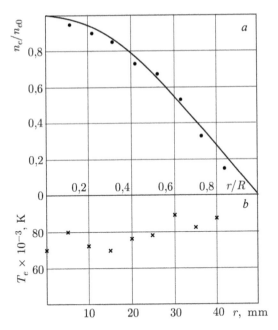

Fig. 3.4. Distribution of the plasma parameters of the RF discharge along the radius of the discharge tube at the centre of the discharge gap (ambipolar diffusion mode). The diameter of the discharge tube is 80 mm, the distance between the electrodes is 60 mm. *a*) distribution $n_e/n_e(0)$ (r/R) (R is the radius of the discharge tube); the experimental points correspond to the conditions: $\omega = 10$ MHz, $V_\sim = 160$ V, $n_e(0) = 1.6 \cdot 10^9$ cm^{-3}; the solid line is the theoretical Schottky distribution [1]: n_e/n_e (0) (r/R) = J_0 (2.4r/R). *b*) the distribution of T_e (*r*).

where R is the radius of the tube, J_0 is the Bessel function of zero order.

The obtained n_e (*r*) and T_e (*r*) distributions are shown in Fig. 3.4.

It can be seen from these curves that the distribution of the electron density along the radius of the discharge tube is close to the Schottky distribution, and the electron temperature in the radial direction varies little.

It was of interest to compare the nature of the $n_e(x)$ and T_e (*x*) distributions from the central region of the discharge toward both electrodes in an asymmetric RF discharge. In this case, the RF potential was applied to only one electrode, and the second electrode remained free, acquiring only the usual constant 'floating' potential, $\varphi_f \sim kT_e/e$, due to contact with the plasma. In this case, the discharge current of RF discharge was closed by a capacitive current to the 'earth', so that it was a single-electrode discharge.

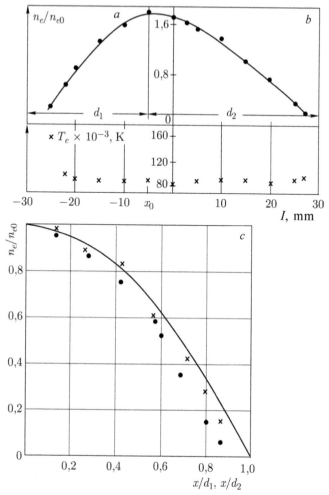

Fig. 3.5. Distribution of plasma parameters along the discharge gap in asymmetric RFCD (ambipolar diffusion regime). *a*) the distribution of $n_e(x)$ for the case of a single-electrode RF discharge ($\omega = 10$ MHz, $V_\sim = 250$ V, $n_e^{max} = 2 \cdot 10^9$ cm^{-3}); active electrode on the left. *b*) the distribution of $T_e(x)$. *c*) Distributions $n_e/n_{e\,max}$ ($x/d_{1,2}$) towards the active and passive electrodes, respectively; The solid line represents the theoretical distribution $n_e/n_{e\,max}$ ($2x/d$) = cos (x / Λ).

The obtained distributions of the electron concentration and temperature towards the active and passive electrodes are shown in Fig. 3.5.

From the curves in Fig. 3.5 it can be seen that in the direction of the active electrode the electron concentration distribution takes a characteristic 'bell' shape, and the temperature T_e significantly changes in the neighbourhood of the electrode. In the direction of the passive electrode, the distribution of n_e is close to the cosine wave, and the electron temperature is almost independent of the coordinate x.

Thus, the experimental data show that at low RF voltages $V_\sim$ the edge effects do not perturb the diffusion distributions $n_e(x)$, $n_e(r)$, where $T_e(x)$, $T_e(r) = $ const in the larger part of the discharge gap.

Estimates for these experiments showed that in this case $f \gg f_{0i}$. As a consequence, the RF field does not penetrate practically into the plasma [203, 204].

In the case of high RF voltages, a significant perturbation of the classical diffusion distribution $n_e(x)$ even in the depth of the discharge gap and a significant inhomogeneity of the $T_e(x)$ distribution are observed (Fig. 3.3).

It is possible to indicate a number of factors related to the boundary conditions that lead to the characteristic 'bell-shaped' distribution of $n_e(x)$: 1) it is estimated that $f_{0i} > f$, and the RF field penetrates deep into the plasma and increases T_e; 2) as will be shown below, high-energy near-electrode electron beams significantly enrich the distribution function of $f_e(\varepsilon)$ by fast electrons. According to the estimates given below for close physical conditions, due to the combined action of the factors of points 1) and 2), the unperturbed coefficient of ambipolar diffusion D_a increases dozens of times ($D_a^B \approx 30D_a$); 3) it was shown in [236] that forced ambipolar diffusion with an effective diffusion coefficient that can significantly exceed the coefficient of ordinary ambipolar diffusion also occurs in an inhomogeneous collisional plasma in a sufficiently strong RF field; 4) for large RF voltage amplitudes, the quasi-stationary potential difference U_s in the NESCL, whose thickness $d_s \sim U_s^{3/4}$ [240], significantly increases; 5) the values of U_s reach values of the order of 10^2–10^3 V in the NESCL, which leads to an accelerated ion transfer to the electrodes and an increase in the charge density gradient ∇n_i at the 'NESCL–plasma' boundary, which increases the diffusion losses of ions.

The last of the explanations given by us is confirmed by the data of [241], in which the dynamics of the ion density distribution $n_i(x, t)$ in the near-electrode region during the RF field period is studied by numerical simulation. Usually it is believed that the inertia of ions is noticeably manifested when $f > f_{0i}$. However, in the above-mentioned work it is shown that in RFCD the inertia of ions manifests itself even at a tenfold increase in the field frequency f by the value f_{0i}

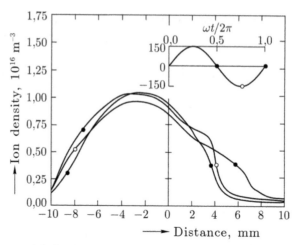

Fig. 3.6. The spatial distribution of the ion density in the phases of the RF field period $\omega t = \pi$, $3\pi/2$, 2π. The grounded electrode is at the point $x = -10$ mm, and the active electrode at the point $x = 10$ mm. Changes in the density in the central part of the discharge are due to a change in the ionization rate necessary for compensation of ion losses [241].

(Fig. 3.6). In this case, it is concluded that there is a sharp decrease in n_i in the vicinity of the electrode, due to intensive stretching ions to the electrode in the negative half-period of the field, which does not have time to replenish in the positive half-cycle of the RF field. This is in accordance with the experimental data shown in Fig. 3.3, 3.5.

Here we should mention the possible impact on the character of the motion of charged particles of average forces [25]

$$F_e = -\nabla \frac{e^2 E_{\sim}^2}{2m(\omega^2 + v_{en}^2)} \quad \text{and} \quad F_i = -\nabla \frac{e^2 E_{\sim}^2}{2M_i(\omega^2 + v_{in}^2)}, \quad \text{arising in regions}$$

with large gradients of the RF field $\nabla E_{\sim}^2$.

However, the estimates made showed that under the experimental conditions the effect of the $F_{e,\ i}$ forces should be substantially weaker than the gas kinetic pressure of the charges $p_e = n_e kT_e$. Moreover, according to our earlier concept, there is not an RF field but a pulsed constant field in the NESCL. Therefore, the appearance in these conditions of the forces $F_{e,i}$ seems, in principle, unlikely.

3.3. Stochastic heating of electrons of near-electrode plasma by an oscillating boundary of a NESCL

At very low gas pressures when the frequency of the RF field is much higher than the collision frequency of electrons with atoms $\omega^2 \gg v_{en}^2$, the power in the RF discharge is significantly higher than that obtained from estimates that take into account only the mentioned collisions.

In the RFCD even in the 'collisionless' case, there is an additional electron heating mechanism due to collisions of plasma electrons with the oscillating outer boundary of the NESCL and called 'stochastic heating'. This physical phenomenon applied to RFCD was first considered in [56].

The essence of the physical mechanism of stochastic heating is compactly stated, in particular, in [6] and consists in the following.

The near-electrode plasma borders on the NESCL, in which the electric field increases most of the period of the RF field from the plasma to the electrode and is directed toward the electrode. This field pushes the electron, which got from the plasma into the layer, back into the plasma.

The boundary between the plasma and the layer oscillates with the velocity u_s. At very low pressures, when collisions of electrons with atoms are rare, we consider the collisionless motion of an electron incident from a plasma in a NESCL with a velocity component v_x normal to the interface. Such acts of penetration of electrons into the layer occur frequently, since the average chaotic velocity of electrons in a plasma is $\bar{v}$ often much larger than the vibrational velocity.

Penetrating into the layer, the electron slows down in the retarding electric field of the space charge. Completely losing the velocity v_x, the electron turns back and, moving already in the accelerating field, returns to the plasma. In the coordinate system associated with the moving boundary of the plasma (potential barrier, wall), the electron undergoes a purely elastic reflection. In the laboratory system, the velocity of the reflected electron is $-v_x + 2u_s$, and its kinetic energy changes as a result of reflection by

$$\Delta\varepsilon = \frac{m(-v_x + 2u_s)^2}{2} - \frac{mv_x^2}{2} = -2mv_x u_s + 2mu_s^2.$$

For certain v_x and $|u_s|$ the electron acquires energy as a result of reflection if the wall moves towards it ($v_x > 0 > u_s$) or overtakes the electron ($u_s > v_x > 0$). If the electron is catching up the wall escaping from it ($v_x > u_s > 0$), it loses energy upon reflection. On average, for two such acts, the electron acquires energy, although in a small amount: $2mu_s^2$.

In general, stochastic heating of plasma electrons is a weak effect, since the velocities of the boundary $|u_s|$ are very small in energy terms.

It should also be noted that, as shown in [249], if the RF discharge is described by a physical model with a uniform distribution of ions in the discharge gap, $n_i(x) = $ const, then the stochastic heating effect of electrons is completely absent.

Moreover, in [6] the possibility of the opposite of the described phenomenon is theoretically predicted. If, in contrast to the consideration of above the conditions, the density of ions in the layer is greater than in the plasma of the central region of the discharge, and the boundary therefore oscillates more slowly than the electrons in the plasma, as a result of reflection from the boundary between the plasma and the layer, it is possible to cool the electron gas rather than heating. According to the same paper [6], there is a form of RFCD, in which the ion density in the NESCL is greater than in the middle of the discharge, but this form is realized at not too low pressures, when stochastic heating is little noticeable against the background of the effect of collisions of electrons with atoms. However, in principle, such an effect of cooling electrons is possible.

3.4. Specific properties of the 'NESCL–plasma' boundary in the RFCD

The physical processes at the boundary separating the regions of the NESCL and the near-electrode plasma are of considerable importance in the mechanism for maintaining the RFCD under investigation and are still poorly understood to this day.

Traditionally, this boundary is considered to be sharp in space and oscillating between the electrode and the location of the maximum thickness of the NESCL [23, 250].

In some works (for example, [213]), the thickness of the given boundary, within which the plasma electron density decreases according to the Boltzmann law, is assumed to be of the order of the Debye radius r_{De}.

In [89], there is only a statement from general considerations that the boundary in question can be spatially unsharp. Accordingly, in the literature there is no analysis of the factors determining the extent and spatial course of the 'NESCL–plasma' section.

Apparently, the existing concepts approximate only to the conditions that arise in the particular case of the RF discharge of the α-type at low RF voltages and field frequencies, when the plasma electrons have a quasi-Maxwellian distribution with a temperature of the order of several eV.

We will consider this issue in a wider range of changes in the parameters of RFCD: voltage $V_\sim$, field frequency f, pressure and kind of gas.

Since the electrode almost always has a lower negative potential with respect to the plasma, the plasma electrons are separated from the electrode by a potential barrier of variable height.

In the case of a wide electronic energy spectrum of a plasma, electrons with different energies penetrate into the potential barrier at different depths. As a result, the interface between the 'plasma–NESCL' regions will not be sharp, as was always assumed in the literature earlier, but will be extended and spatially structured.

The structure of the boundary depends on the nature of the plasma EES. The greatest depth of penetration into the region of the potential barrier is determined by the maximum value of the energy of the plasma electrons.

As a result of interaction of a broad non-Maxwellian EES of the plasma having a non-monotonic course in the high-energy region with the potential barrier in the NESCL, potential wells (we call their 'lacunae') filled with electrons should form.

Plasma electrons that are incident on the potential barrier of the NESCL stop in their motion toward the electrode (along the x axis) in the barrier regions, where the x-th component of their kinetic energy is compared with the energy of the electric field that retards the electrons. The spatial distribution of the density of electrons trapped by the field at the plasma–NESCL boundary is due to the potential barrier profile $V(x)$ and the nature of the plasma EES. Therefore, the 'plasma–NESCL' boundary in the case of a wide EES of the plasma must be extended in space and have a structure reflecting the energy distribution of the electrons of the near-electrode plasma.

The spatial variation of the potential $V(x)$ in the NESCL is due to the corresponding distribution of the excess charge density of positive ions.

In the process of deceleration and stopping of electrons incident on the barrier in a certain spatial region of the NESCL, a seed group of electrons is formed in this region, continuously replenished by a stream of fast plasma electrons from the energy interval of the non-monotonicity of the EES. In this case, the local excess density of positive ions decreases, and local spatial sagging of the

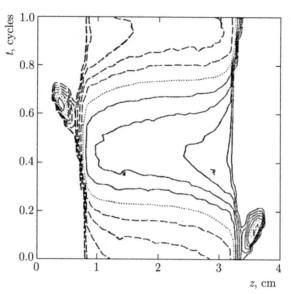

Fig. 3.7. Spatio-temporal distribution of the electric field in the RFCD. [251].

potential – a potential well – arises. The instantaneous location and parameters of the 'lacunae' (spatial configuration, depth of the potential well) are determined by the physical conditions in the RFCD under investigation.

Confirmation of the hypothesis advanced by the author about the formation in the RFCD of the spatially extended, structured 'plasma–NESCL' boundary, which includes the resulting potential wells, can be seen in the theoretical paper [251]. In the paper cited on the self-consistent kinetic calculations of RFCD in He, the local triple reversal of the direction of the field inside the contracting NSCH is established, which means the formation of a potential well for electrons (Fig. 3.7). We note that in the paper [251] nothing is said about potential wells.

In this case, as can be seen from the results of [251], the instant of the appearance of a potential well coincides with the arrival of a high-energy group of electrons. This group occurs before the expanding NESCL and passes the central region of the plasma to the opposite electrode. [251] also shows that these high-energy electrons are not γ-electrons, since the picture does not change qualitatively even in the calculations assuming the absence of emitted electrons ($\gamma = 0$).

However, in our case, the beam group consists precisely of γ-electrons, since our RF field frequency was an order of magnitude lower, and the expanding NESCL could not form such a high-energy group of electrons.

3.5. The possibility of a discrete mechanism for transport of plasma electrons to the electrodes of the RFCD

For RFCD modes with high values of $V_{\sim}$ and f, when the plasma does not have time to diffuse to the electrode in a time equal to the half-period of the RF field, the question arises of the mechanism of the transport of plasma electrons to the electrodes, which ensures the existence of a stationary discharge.

The observation of a real-life stationary RFCD under similar conditions can be explained as follows.

The oscillogram of charge flows to the electrode for a discharge in Xe reveals a very sharp leading edge of the electron pulse (Fig. 3.8).

The electron energy spectrum of a plasma can not have such an abrupt change, except for a narrow energy interval including a quasi-monoenergetic electron beam.

However, the latter has a significantly lower density, compared with the plasma density.

More naturally, the foregoing sharp front of the oscillogram is associated with the arrival of a spatially compact electronic formation constrained by the steep wall of the potential well.

Let us turn again to the hypothesis of the formation of potential holes ('lacunae') put forward earlier at the boundary of the NESCL at the time of its maximum width.

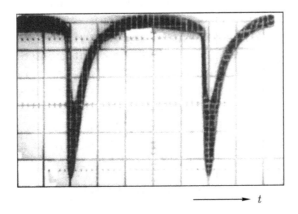

t

Fig. 3.8. Oscillogram of charged particle fluxes to the electrode (symmetrical discharge). Xe, $p = 0.1$ Torr, $f = 1.2$ MHz, $V_{\sim} = 1000$ V.

A formed 'lacuna' is filled with electrons, able to fly to it and stay in it.

Within each period of the RF field, when the height of the near-electrode potential barrier is lowered, the fastest plasma electrons shift the front boundary of the potential well deep into the NESCL: the 'lacuna' begins to move toward the electrode. The fastest electrons of plasma EES catch up with the 'lacuna' and fill it. With the minimum height of the potential barrier, the potential well comes into contact with the surface of the electrode and splashes the electron charge brought on it.

Then, the potential barrier of the NESCL starts to recover again, and the arrival of electrons to the electrode relatively smoothly stops. On the oscillogram, we see that the trailing edge of it is more gentle (Fig. 3.8).

It is obvious that the average velocity of the 'lacuna' displacement is established self-consistently with the dynamics of the near-electrode potential barrier and is equal to

$$u_l = \frac{d_s}{T/2} = \frac{2d_s}{T}.$$

For the conditions of this experiment:

$$u_l \geq 2 \cdot 10^6 \text{ cm/s}.$$

Thus, even in the RFCD mode with the low temperature of the main electron mass of the near-electrode plasma, a sufficient self-consistent number of electrons can be transferred to the electrode in potential wells for each period of the RF field.

Thus, depending on the physical parameters of the RFCD, transport of plasma electrons to electrodes providing a quasi-stationary discharge state can occur by two mechanisms: 1) traditional, continuous – by diffusion of plasma to the electrode during a positive half-period of the field; or 2) by the proposed new, discrete transport mechanism with the help of potential wells periodically intersecting the NESCL.

3.6. The edge resonance effect in the RFCD

The edge resonance effect in the RFCD is of interest, which arises as a result of the interaction of the NESCL with a near-electrode plasma. This effect was experimentally investigated in [188], where the resonant behaviour of low-pressure RFCDs is considered, which is associated with a successive oscillatory circuit of the 'plasma–NESCL' system.

The physical nature of this resonance can be represented on the basis of the linear theory of a cold homogeneous plasma. In this case, the plasma can be regarded as an equivalent parallel electric circuit with inductance L_p and a capacitance C_p:

$$L_p \cong m d_p / n_e e^2 S \quad \text{and} \quad C_p \cong S / 4\pi d_p,$$

where S is the electrode area, d_p is the plasma length. The plasma characteristics L_p and C_p give a parallel resonance at the Langmuir frequency ω_{e0}. At frequencies lower than the electron Langmuir frequency, the plasma behaves as an inductance and resonates with the capacitance of the layer $C_s \cong S/4\pi d_s$, where d_s is the width of the layer. The frequency of the consecutive 'plasma–layer' resonance is

$$\omega_r = \frac{1}{\sqrt{L_p C_s}} = \omega_{e0} \sqrt{\frac{d_s}{d_s + d_p}}.$$

In the case when the operating frequency $\omega = \omega_r$, a series resonance can be observed as a minimum on the I–V characteristic of the RFCD and a change in the sign of the phase shift between the discharge current and voltage [188].

Such resonance discharges have been studied for scientific and applied purposes [252–254].

Modern experimental and industrial HF systems operate as a rule with high currents and plasma densities, when the frequency ω_r is much larger than the operating frequency ω. In this case, the resonant potential fluctuations can be excited by various factors: the external source of the RF field, the noise or harmonics of the operating frequency ω, generated inside the RFCD itself.

As mentioned earlier, RFCD is an explicitly non-linear system. Even when an ideally sinusoidal voltage is applied to the electrodes, the harmonics of the operating frequency appear in the fluctuations of the plasma potential and the discharge current. The non-linear properties of the NESCL produce these harmonics. In this case, the amplitudes of the harmonics decrease with increasing ordinal number of the harmonic. Therefore, only the first few harmonics of the operating frequency have amplitudes sufficient for their recording by electrical diagnostic methods.

It should be noted, however, that it was found in [73, 187, 255] that the discrete spectrum of the operating voltage harmonics may contain some selected harmonic with an unexpectedly high amplitude. The physical nature of this phenomenon is explained in [255] as follows. The low-pressure RFCD, as a non-linear system, can be represented using an equivalent electrical circuit, shown in Fig. 3.9.

Capacitances of near-electrode layers of RFCD can be divided into a variable in time and a constant part. The variable in time component of the capacity of the layers produces non-linearity, while the constant part (independent of the phase of the operating voltage) resonates with the inductance of the plasma. Although the resonance itself is not directly related to the harmonics of the operating frequency ω, significant fluctuations can be excited at the harmonic of this frequency, which may be closest to the resonant frequency ω_r.

Experimental determination of the resonant frequency

$$\omega_r = \omega_{e0}\left(\frac{d_s}{d_s + d_p}\right)^{1/2}$$

gives the value of the plasma frequency ω_{e0}, if the width of the layers d_s is known. Knowing the value of ω_{e0}, we can calculate the electron density n_e. Thus, this method of studying the resonance 'plasma-layer' can be considered as a method for diagnosing plasma RFCD.

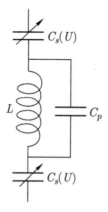

Fig. 3.9. Equivalent electric circuit of low-pressure RF capacitive discharge. L_p is the plasma inductance, C_p is the plasma capacitance, C_s (U) is the nonlinear capacitance of the near-electrode layer [188].

3.7. The edge effect of the RF field in the RFCD

As mentioned earlier, another edge effect in RFCD is the edge effect, which is expressed in the output of a part of the RF field strength lines in the near-electrode region from the discharge gap in the radial direction.

The edge effect under the conditions of RFCD was experimentally investigated in [189].

The electrodes are of limited dimensions, especially the active small diameter electrode in an asymmetric discharge. From the sharp boundaries of the electrodes in the surrounding space, the lines of the potential electric RF field diverge, closing to the 'ground'.

The influence of always present parasitic capacitances in the RFCD the traditional planar geometry is further increased at higher field frequencies due to the reduction of capacitive impedances in the direction of the 'ground'.

The spatial extent of manifestation of the edge effect depends, generally speaking, on all the RFCD parameters and, first of all, on the geometry of the discharge gap.

Below we describe in more detail one of the main physical influences of the edge effect: the appearance of electron beams in the direction from the dielectric walls to the axis of the discharge tube (the wall electron beams – WEB).

The developed edge effect associated with the exit of a portion of the electric field lines from the discharge gap leads to additional energy losses in the RFCD.

However, the outgoing HF field, whose lines of force cross the space charge layer near the side wall of the discharge tube, cause the effect of 'RF detection' in this layer [28]. In this case, the potential difference U_s, which depends on the parameters of the RF field, arises in the wall layer, which can greatly exceed that due to a simply 'floating potential'. Thus, there are all the conditions for the emergence of wall electron beams (WEB), which return an appreciable fraction of the outgoing energy of the RF field back to the discharge gap.

On the quantitative side of the edge effect from the energy point of view, one can judge approximately by the parameters of the WEB presented below.

Here it should be noted that the edge effect is almost completely eliminated in the case of using the coaxial geometry of the discharge gap of the RFCD.

3.8. Penetration depth of the electric field in the near-electrode plasma of the RFCD

In connection with the properties of the electric field discussed earlier in the RFCD, the question arises as to the penetration depth of the electric field in a near-electrode plasma of the RFCD. In this case, of course, we are talking about a γ-type RFCD, in the near-electrode layers of which the resulting electric field takes the form of unipolar pulses of a quasi-stationary field, which follow with the frequency of the RF field supporting the discharge. This kind of resultant field arises from the superposition of the alternating field from an external RF power source and a quasi-stationary space charge field in a layer whose voltage amplitudes in a γ-discharge are of the same order (Fig. 2.21).

In the α-type RFCD with reduced voltages, the RF field dominates the quasi-stationary RF field, and the process of interaction of the RF field with the operating frequency ω with the near-electrode plasma occurs in the usual way [277].

In the case of a γ-type RFCD, there are unipolar pulses of a quasi-stationary electric field in the NESCL, and only the duration of these pulses, which is about half cycle-long, depends on the frequency of the RF field ω:

$$\tau_i \sim \frac{T}{2} = \frac{\pi}{\omega}.$$

In this case, the intensity vector of the resultant field $\mathbf{E}$ in the NESCL has a constant direction, varying only in amplitude during the period of the RF field.

Analyzing the results of previous measurements by R. Gottscho [185] of the spatial distribution of the strength of the RF field $E_\sim(x)$ in the near-electrode region of RFCD, it can be stated that the penetratioin depth of the RF field in a wide range does not depend on the frequency of the field (Fig. 1.37).

Beyond the limits of the NESCL, in which a superposition of RF and quasi-stationary fields occurs, the residual purely RF field is already weak, and its attenuation is difficult to observe. Apparently, in the first approximation, the penetration depth of the electric field at the NESCL boundary of the γ-type RFCD is of the order of the electron Debye radius $\delta \sim r_{De}$, as in the case of the electrostatic field.

Thus, the described phenomenon of the superposition of the RF field and the quasi-stationary electric field represents another edge effect, leading to a specific character of the phenomenon of the attenuation of an alternating electric field in the near-electrode region of the RFCD.

4

Near-electrode electron beams in RF capacitive discharge

Representations about the role of boundary surfaces in the RFCD mechanism have undergone a certain evolution.

At first, it was believed that phenomena at the boundaries do not at all affect the properties of the RF discharge [1]. Next, a classification was proposed for the varieties of the RF discharge [23] – the α-discharge and the γ-discharge, where in the latter case, the significant role of the electrons emitted by the electrodes in maintaining the discharge was assumed.

As previously reported, for the first time the near-electrode electron beams in the low-pressure RFCD were experimentally discovered by the author in 1968 [47, 48, 50, 51].

Later, it was suggested and experimentally confirmed the existence of electron beams from the dielectric walls of the discharge tube [189].

It should be noted that the appearance of electron beams is one of the most important edge effects due to the presence of the space charge layer (SCL) near the surfaces of the discharge gap.

At the same time, in the literature there is practically no data on the experimental study of electron beams in RFCD, with the exception of papers [218, 219, 242], which confirmed their existence.

Directional electron fluxes are a characteristic attribute of low-pressure RFCD. Moreover, the motion of these fluxes can be oriented not only from the boundaries into the depth of the discharge gap. Thus, in the specific modes of RFCD, which were not of primary interest, volumetric plasma structures were observed that emitted electron beams toward surfaces that bound the discharge space [243].

Existing practically the whole period of the RF field, the beams are essentially modulated in time with respect to energy and density, which leads to modulation of the parameters of the near-electrode plasma with the frequency of the applied field.

In carrying out these studies, it is important to take into account the dependence on all parameters of RFCD as parameters of the beams themselves, and the effects of interaction of beams with the plasma.

To avoid superposition of the effects of opposing beams from opposite electrodes, as well as to increase the parameters of the beam under investigation, it was necessary to use a highly asymmetric or single-electrode RFCD in a number of experiments.

4.1. The phenomenon of the appearance of boundary electron beams in RFCD

4.1.1. Near-electrode electron beams

In the vicinity of the RFCD electrodes there are constantly conditions for the appearance of electron fluxes directed toward the centre of the discharge gap.

Indeed, the potential of the plasma is always higher than the electrode potential [207], and factors that cause electronic emission from the electrodes as a result of the arrival of ion fluxes, photons, excited neutral particles, etc., act continuously.

Near-electrode electron beams at RFCD were detected by the probe method [47, 48]. The corresponding experiments are described in Chapter 2, where the results are also presented (Figs. 2.28, 2.29).

A significant degree of anisotropy of the motion of electrons in the near-electrode region of the RFCD is also indicated by the difference in the type of the current–voltage characteristic of a plane probe with its working surface facing the electrode and in the opposite direction (Fig. 4.1).

The anisotropic nature of these probe curves can be explained by the action of fast electrons, which cause additional emission of electrons from the surface of the probe.

The maximum decrease in the probe current was observed when the surface of the probe was facing a stream of fast electrons (curve *1*). When the plane of the probe is parallel to the direction of the flow (curve *2*), the emission is caused by fast scattered electrons and is weaker.

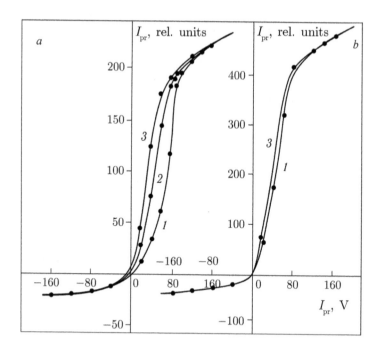

Fig. 4.1. Dependence of the probe characteristics on the orientation of the flat probe in the RF discharge (ω = 7 MHz, $V_{\sim}$ = 280 V). *a*) near-electrode region (distance from probe to electrode d_{pe} = 10 mm). *1* – the surface of the probe faces the electrode, *2* – the surface of the probe is parallel to the axis of the discharge tube, *3* – the probe faces the electrode with an isolated side. *b*) The central region of the discharge (d_{ape} = 30 mm, the designations are the same).

Curves *1, 2, 3* practically coincide in the regions of saturation of the ion and electron currents. In the region of the ion saturation current, large potentials negative relative to the plasma are fed to the probe. Therefore, very few electrons get to the probe, and the emission caused by them is small. In the region of the electron saturation current (behind the potential of the plasma space), the emitted electrons must return back to the probe.

An effective way to detect the directional motion of electrons in the near-electrode region of RFCD is the action of a homogeneous permanent magnetic field $\mathbf{H}_0$ on the discharge.

The change in the luminescence pattern of the RFCD as a result of the action of the magnetic field $\mathbf{H}_0$ produced by the Helmholtz coils is shown in Fig. 2.31. As can be seen, the deflection of the luminescence at the RFCD electrodes is directed in opposite directions, which corresponds to the directions of the motion of the opposing electron beams.

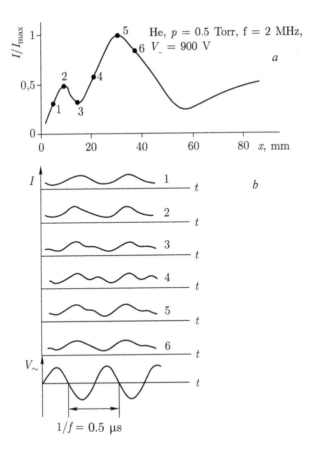

Fig. 4.2. The oscillograms of the luminescence intensity I (t) at different distances from the electrode corresponding to the indicated points of the quasi-stationary distribution I (t).

A comparison of the experimental results obtained for RFCD and DC glow discharge (DCGD) indicate the identical nature of the effect of the field H_0 on the luminescence in the near-electrode regions of these discharges. From this we can conclude that the motion of the glowing electrons in the region of the cathode dip of the DCGD is identical, and in the near-electrode region of the RFCD, the unidirectional motion of electrons from the electrode.

An additional confirmation of what has been said above is the established identity of the quasi-stationary near-electrode distributions of the integral luminescence I (x) in the RFCD and DCGD (Figs. 1.33, 1.34).

The previously presented model ideas on the behaviour of the luminescence in the near-electrode region of the γ-type RFCD presuppose an asymmetric oscillation of the luminescence at the second harmonic of the RF field (Fig. 2.21), are confirmed by our observations and oscillograms of the luminescence in [27, 218].

More detailed information on this question is given by the conducted studies of the luminescence behaviour in time in the characteristic regions of the quasi-stationary distribution $I(x)$, analogous to the regions of 'cathode luminescence', Crookes dark space, 'negative luminescence' of DCGD (Fig. 4.2, points 1–6). In this case, the oscillogram of the luminescence $I(t, x)$ was recorded simultaneously with the RF potential at the electrode, so that there was a connection of the oscillogram of the luminescence to the phases of the RF field period.

According to the accepted analogy of the spatial regions of the RFCD and DCGD, it is believed that the observation points 1–4 are in the NESCL of the RFCD. The resulted oscillograms of the luminescence show a significant asymmetry in the half-periods of the RF field: in the negative half-cycle there is intense luminescence, in the positive half it is weaker. At the observation point 1 closest to the electrode, only one luminescence maximum is displayed on the oscillogram in the negative half-cycle of the RF field, i.e., at this point the luminescence oscillates with the frequency of the RF field f. When moving away from the electrode (points 2–4), a second substantially lower maximum of luminescence appears in the positive half-period, which, after passing the maximum value of the amplitude, decreases at large distances from the electrode when approaching the quasi-neutral boundary of the near-electrode layer.

In connection with the constant presence of electron emission from the electrode and the voltage pulse accelerating the emitted electrons into the negative half-cycle of the RF field, the appearance of a maximum of the luminescence intensity in this half-period can be attributed to the excitation of the gas by a pulsed electron beam from the electrode. A pulse amplification of the luminescence intensity once during the RF field was observed not only in the near-electrode region, but also in the plasma volume in the neighbourhood of the electrode. An analogous picture shifted in phase by π was observed at the opposite electrode.

The appearance on the oscillograms corresponding to observation points 2–4, some increase in the luminescence in the positive half-period of the field, is apparently due to the diffusion approach of the plasma boundary to the observation points. We recall that in the positive half-cycle of the RF field the potential barrier blocking the plasma electrons falls off.

For the conditions under consideration, the distance from the quasi-stationary boundary of the layer (point 4) to the observation point 3 is z = 5–7 mm (Fig. 4.2).

Analysis of the oscillograms shows that the most complex character is that which corresponds to the observation points between the first and second maxima of the distribution $I(x)$, that is, between analogues of 'cathode luminescence' and 'negative luminescence' of the DCGD.

Bearing in mind that in the luminescence structure of the DCGD the cathode drop edge lies between the first and second maxima, we can expect at the same place the time-averaged position of the boundary of the near-electrode layer of the RFCD. Since in the RFCD the position of this boundary oscillates in time, a complicated picture of the oscillation of the luminescence (oscillograms 3–4) should be expected in this region of space.

Thus, using this experimental method, we can find the region of space in which the boundary oscillation of the near-electrode region of RFCD is localized.

We estimate the amplitude of the displacement of the boundary of the layer Δz under the experimental conditions: He, p = 0.2 Torr, f = 4 MHz, T_e = 1 eV, the collision frequency of electrons with atoms v_{en} = 3 · 10^8 s^{-1}, electron diffusion coefficient D_e = 6 · 10^6 cm²/s, the half-period of the RF field $T/2$ = 1.25 · 10^{-7} s.

$$\Delta z \sim \left(\frac{T}{2}D_e\right)^{1/2} = \left(1.25 \cdot 10^{-7} \cdot 6 \cdot 10^6\right)^{1/2} = 0.9 \text{ cm}$$

i.e. the oscillating boundary of the layer must pass through the observation point.

A weak increase in the luminescence in the positive half-period can not be, apparently, due to heating by the RF field, since it has a character that is essentially localized in space – between points 2–4 directly outside of which the oscillograms of the luminescence have one maximum during the RF field period.

For the conditions corresponding to Fig. 4.2, we estimate the average residence time t_e in the layer of emitted electrons. Let the field strength in the layer be $E \sim$ 300 V/cm, the layer thickness $d_s \sim$ 1 cm, p = 0.2 Torr, P_1 = 10 cm^{-1} · Torr^{-1} [244].

The velocity of the electrons $v_e = b_e E$, $b_e = \dfrac{e}{m v_{en}}$; $v_{en} = pP_1 v_e$;

$$v_e = \frac{eE}{mpP_1 v_e} \qquad v_e = \left(\frac{eE}{mpP_1}\right)^{1/2} = \left(\frac{4.8 \cdot 10^{-10} \cdot 300 \cdot 2}{300 \cdot 9.1 \cdot 10^{-28} \cdot 0.2 \cdot 10}\right)^{1/2} = 5 \cdot 10^8 \text{ cm/s}$$

$$;$$

$$t_e \approx \frac{d_s}{v_e} = \frac{1}{5 \cdot 10^8} \approx 2 \cdot 10^{-9} \text{ s}$$

1

Thus, $t_e \ll T$, that is, the emitted electrons do not participate in the oscillatory motion in the layer.

Observations of the behaviour of radiation in time at different distances from the electrode were also carried out for individual spectral lines. At the same time, the optical signal from the output slit of the MDR-12 monochromator came to the FEU-38 photomultiplier, whose output was connected to the input of a stroboscopic oscilloscope S7-13. A two-coordinate recorder LKD-4 was connected to the analog output of the oscilloscope.

In these measurements, the frequency of the RF field was chosen to be sufficiently low ($f = 2.4$ MHz), so that the relation $\tau_{ex} < 1/f$ for the most intense spectral lines, where τ_{ex} is the lifetime of the atom in the excited state corresponding to the upper level of the optical transition under study.

Two types of measurements were made: obtaining 'time' and 'phase' radiation characteristics of RFCD.

When recording the 'time' characteristics (oscillograms), the position of the tube was fixed, and the strobe moved smoothly within 1–2 periods of the RF field on the oscilloscope screen. These characteristics were taken for different distances from the electrode to the observation point.

When the 'phase' characteristics were obtained, the strobe of the oscilloscope was set in series in four characteristic phases of the RF field period, and the discharge tube moved before the input of the diaphragm system.

The oscillograms of the emission of the spectral lines He I = 7065 Å, Ne I = 6402 Å and Ne I = 7032 Å, obtained using a stroboscopic oscillograph, are shown in Figs. 4.3–4.5 respectively.

As follows from the adopted model of RFCD, the maximum energy of the beam electrons, and, consequently, the maximum amplitude of the luminescence oscillations should be achieved in phase 1 of the RF field. Similar cases occurred (for example, Fig. 4.5).

However, in the vast majority of cases, the maximum of the luminescence oscillations preceded phase 1 (Figs. 4.3 and 4.4). A number of experiments indicate the dependence of the shift of the

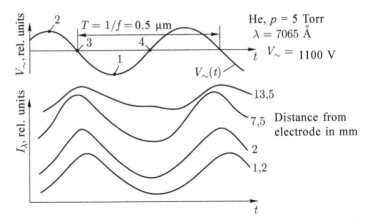

Fig. 4.3. The oscillograms of the intensity of the spectral line He, $I = 7065$ Å, at different distances from the electrode in the RFCD.

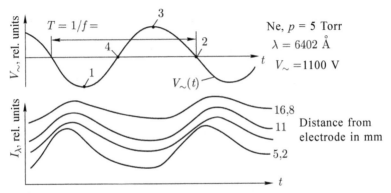

Fig. 4.4. The oscillograms of the intensity of the spectral line Ne, $I = 6402$ Å, at different distances from the electrode in the RFCD.

maximum of the luminescence oscillations relative to phase 1 on the gas pressure, the shift increases with increasing pressure.

It is known that in RFCD with increasing pressure, under other fixed conditions, in the investigated pressure range, the concentration of charged plasma particles increases.

As the negative potential of the electrode increases, during one period of the RF field the beam energy and the intensity of the luminescence that it produces increase. At the same time, the emission processes from the electrode (bombardment by ions, excited particles, photoemission) are intensified, reducing the negative potential of the electrode. Apparently, therefore, the maximum of the luminescence intensity is observed somewhat earlier than phase 1, where because of the emission processes the value of the negative potential will no longer be the maximum in magnitude.

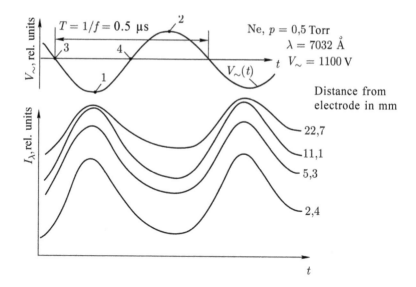

Fig. 4.5. The oscillograms of the intensity of the spectral line Ne, $I = 7032$ Å, at different distances from the electrode in the RFCD.

Just as in the case of oscillograms of the integral luminescence $I(t, x)$ (Fig. 4.2), the oscillograms of the emission of the spectral lines $I_\lambda(t, x)$ shown above (Figs. 4.3–4.5) show a tendency at certain distances from the electrode to increase the radiation intensity in the positive half-cycle of the RF field (a protracted trailing edge of the oscillogram pulses).

Since on all the oscillograms $I(t, x)$, $I_\lambda(t, x)$ there is no increase in luminescence in the positive half-period of the field near the electrode surface, then for the investigated physical conditions in the RFCD it should be concluded that the plasma does not have time to shift to the electrode surface during this half-period.

The studies described above contain information on the behaviour of the radiation intensity at individual points of the near-electrode region of the discharge.

A spatio-temporal picture of the luminescence in the entire near-electrode region of the RFCD during several periods of the RF field, obtained with the help of ultrafast photography by the camera LV-03 ('time magnifier') with an exposure time $\tau = 10^{-6}$ s, is shown in Fig. 2.22.

It is evident that the luminescence in the near-electrode region: 1) has a pulsed character with the frequency of the RF field; 2) lasts during the negative half-cycle of the RF field; 3) extends from the

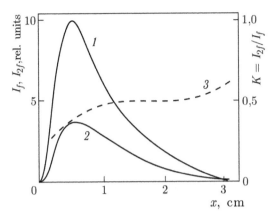

Fig. 4.6. The spatial distributions of the amplitudes of the luminescence oscillations $I_f(x)$ and $I_{2f}(x)$ and the ratio $K(x) = I_{2f}/I_f$, He, $p = 2.2$ Torr, $f = 1.2$ MHz, $V_\sim = 750$ V. $1 - I_f(x)$, $2 - I_{2f}(x)$, $3 - K(x)$.

electrode to the centre of the discharge gap, monotonically decreasing in intensity.

The established character of the luminescence unambiguously confirms the hypothesis advanced about the existence of pulsed near-electrode electron beams in the RFCD.

We also note that an analysis of the character of the glow in Fig. 2.22 shows: 1) outside the NESCL, the plasma glow is modulated at a frequency of $2f$;

2) the glow of the plasma does not approach close to the electrode during the entire period of the RF field.

The amplitude of the luminescence at the frequency of the RF field $I_f(x)$ at any point in the discharge gap characterizes the local intensity of excitation of the gas by the particles of the electron beam and makes it possible to diagnose its behaviour in the discharge.

The explanation of the luminescence oscillations at a doubled frequency $2f$ in different regions of the RFCD is not unambiguous. Indeed, the appearance of the second harmonic of the signal can be due to: 1) heating of electrons by a local RF field; 2) by oscillation of the NESCL boundary and 3) by overlapping in the space of electron beams emerging from the opposite electrodes in the antiphase.

A typical example of the distributions of $I_f(x)$ and $I_{2f}(x)$ in the near-electrode region of RFCD is shown in Fig. 4.6. It can be seen that near the electrode the oscillation amplitude at the field frequency is much greater than the amplitude of the second harmonic, sometimes within two orders of magnitude.

According to the foregoing, in the NESCL of the γ-type RFCD under investigation, there is practically no alternating RF field appearing beyond the boundary of the near-electrode layer. This is confirmed by the obtained distributions of $I_f(x)$ and $I_{2f}(x)$ of the type shown in Fig. 4.6, which shows that as the distance from the electrode, the relative contribution of the RF field to the excitation of the gas $K(x) = \dfrac{I_{2f}(x)}{I_f(x)}$ increases.

It should be noted that there will be almost no oscillations at frequency $2f$ if condition $2\omega > \delta_e \nu_{en}$ ($\omega = 2\pi f$) is satisfied, δ_e is the fraction of energy transferred by the electron in a collision. For the case shown in Fig. 4.6:

RFCD in He, $\nu_{en} = 5.3 \cdot 10^9$ s^{-1}; $\delta_e = 3 \cdot 10^{-4}$; $\omega = 7.5 \cdot 10^6$ s^{-1}; $\delta_e \nu_{en} = 1.6 \cdot 10^6$ s^{-1}, i.e. $2\omega > \delta_e \nu_{en}$, and $\omega \geq \delta_e \nu_{en}$. Thus, the modulation of the beam emission should be manifested more strongly, which is observed experimentally.

At the same time, even if the condition $2\omega < \delta_e \nu_{en}$ is satisfied in the of the γ-type RFCD region, the amplitude of the oscillations at the frequency f exceeded that at the frequency $2f$ in all the experimental conditions studied.

The nature of the motion of electrons in the near-electrode region of the RFCD was also investigated using the method of polarization spectroscopy developed by the group of S.A. Kazantsev at the Leningrad State University [216, 245]. The intensity distribution of the spectral lines $I_\lambda(x)$ along the discharge axis and the degree of polarization of radiation $P_\lambda(x)$ were studied and the general character of luminescence was recorded (Figs. 2.33 and 2.34).

For all the spectral lines studied, the distribution $I_\lambda(x)$ was represented by a curve increasing from the electrode to the centre with a maximum at a certain distance, after which there was a monotonous decrease in the central region of the discharge. At the minimum used pressures, the above-mentioned near-electrode maximum disappeared, and the dependence $I_\lambda(x)$ was a curve monotonically increasing toward the centre of the discharge.

A pronounced polarization maximum was recorded in the obtained distribution of the degree of polarization $P_\lambda(x)$. The maximum was was located closer to the electrode than the radiation intensity maximum of the curve $I_\lambda(x)$ (Fig. 2.33). Moreover, even in cases when the distribution $I_\lambda(x)$ did not exhibit a near-electrode maximum, the maximum of the degree of polarization was clearly recorded near the electrode (Fig. 2.34).

A comparison of the polarization curves $P_\lambda(x)$ with the character of the integral luminescence gives additional information. The brightest region of integral luminescence begins immediately at the boundary of the near-electrode dark space. The application to the discharge volume of a weak magnetic field made it possible to establish that the region of the brightest emission is a consequence of the excitation of a gas by electron beams. Localization of the near-electrode maximum of the curve $P_\lambda(x)$ on the boundary of the dark space allows us to assume that the formation of the electron beam terminates in the boundary section of the NESCL.

As is well known [182], the polarization of radiation in a discharge indicates the existence of directed electron fluxes, which is an additional confirmation of the existence of near-electrode electron beams in the RFCD studied.

Significantly later than the moment of detection of near-electrode beams in RFCD, one more experimental confirmation of their existence by an original method was carried out in [219]. In this case, celluloid films that are opaque for electron beams were installed before the electrodes of the RFCD. The time-resolved photography of luminescence showed the presence of rectilinear geometric shadows behind these films in the negative half-cycle of the RF field at the electrode. The latter gave grounds for assuming that in the near-electrode region of RFCD at the appropriate instants of time the electrons move not vibrationally or chaotically, but along 'ballistic trajectories' directed from the electrode to the centre of the discharge.

As an indirect additional confirmation of the existence of near-electrode beams, two more experimental facts can be found.

The first of them consists in the detection in the optical spectra of the plasma of RFCD of intense 'ion' lines, which require for their excitation electrons of sufficiently high energies (for example, for the line He II 4685Å $\varepsilon_e = 75.6$ eV). The intensity of these lines was maximal in the neighbourhood of the electrodes and decreased monotonically to the centre of the discharge.

As a second fact, we will consider the observed practical absence of a gradient of the quasi-stationary potential in the short discharge gap of the RFCD (Fig. 2.5). The latter can be interpreted as overlapping at the centre of two near-electrode regions – analogues of the 'negative glow' region of DCGD, which is characterized by a very small potential gradient and anomalously low values of the electron temperature and generated by electron beams from the region of the cathode drop [151]. The measurements we made in this area of RFCD also yielded values of $0.05 \leq T_e \leq 1$ eV.

As a result, on the basis of a number of various experimental studies it has been established that in low- and middle-pressure RFCDs near-electrode electron beams are present, the parameters of which depend on all discharge characteristics and will be considered below.

The mechanism of formation of near-electrode beams is as follows. During the whole period of the RF field, electrons (γ-electrons) are emitted from a surface of the electrode under the influence of a number of mechanisms. Between the electrode and the plasma there is always an accelerating γ-electron field, reaching the maximum stresses in the negative half-cycle of the RF field. Often, the emitted electrons pass through the NESCL in a virtually collisionless regime and gain energy corresponding to the potential difference in the near-electrode layer. The time for the passage by the γ-electrons of the NESCL is much shorter than the duration of the RF field period. At the outer boundary of the NESCL, the formation of electron beams injected into the near-electrode plasma is completed.

The external boundary of the NESCL oscillating with time transfers the additional velocity to the 'plasma' electrons that are in contact with it ('stochastic heating' [56]); this velocity is an order of magnitude smaller than the velocities of the electron beams.

Quantitative estimates characterizing the formation of electron beams in the NESCL are given below.

4.1.2. Wall electron beams

In the course of research, a hypothesis has been put forward that there exist not only near-electrode electron beams but also the entire surface of the cavity surrounding a sufficiently short discharge gap – the dielectric walls of the discharge tube – and can emit electron beams into the discharge volume.

Weak electron beams should also be from the walls of the discharge tubes running parallel to the plasma of the positive column in the DCGD. The energy of these beams can not exceed a certain value

$$\varepsilon_{ewb} < \varepsilon_{e0} + \varepsilon_{efl},$$

where ε_{e0} = 4–12 eV [225] is the energy of the emitted electrons, $\varepsilon_{efl} = e\dfrac{kT_e}{2e}\ln\dfrac{M_i}{m} = \dfrac{kT_e}{2}\ln\dfrac{M_i}{m}$ is the energy acquired by the emitted electrons in the wall SCL in the field of the 'floating' potential. The energies of such wall electron beams (WEB) do not exceed several tens of eV.

The situation is different in the RFCD, where there is always a previously mentioned edge effect of the RF field in the vicinity of the discharge electrodes. As a result of this effect, a radial component of the RF field applied to the electrodes appears, passing through the wall SCLs in the surrounding space. Here, as in the NESCL, there will also be all the necessary conditions for the manifestation of the RF detection effect [28].

Owing to the above-mentioned reasons, an additional quasi-stationary potential difference V_0' must appear in the wall space charge layers which at appropriate parameters of RFCD can repeatedly exceed the voltage of the 'floating' potential, which is confirmed experimentally below.

Thus, the electrons emitted from the walls will have energies on the outer boundary of the SCL

$$\varepsilon_{ewb} \leq \varepsilon_{e0} + \varepsilon_{efl} + eV_0'.$$

Naturally, as the distance from the electrodes increases, the values of eV_0' will decrease as a result of the exit from the zone of the edge effect.

The experimental scheme by which the WEBs were detected and investigated is shown in Fig. 4.7.

In this case, the presence of an electron beam from the wall was identified with the presence of luminescence oscillations at the frequency of the RF field, in connection with which the corresponding radial distributions $I_f(r)$ were obtained in the discharge tube.

The discharge tube was a quartz cylinder 7 cm long and 6 cm in diameter, the ends of which were covered with transparent quartz plates.

In this tube, one-electrode RFCD was excited, and the electrode was external and consisted of metal rods 0.5 cm in diameter, 2 cm and 6.4 cm in diameter, and its ends which were tightly pressed against the end plate of the tube.

The free top of the discharge tube was used as a window for optical observations, through which the distributions of $I_f(r)$ were obtained. To do this, light from the discharge through a diaphragm

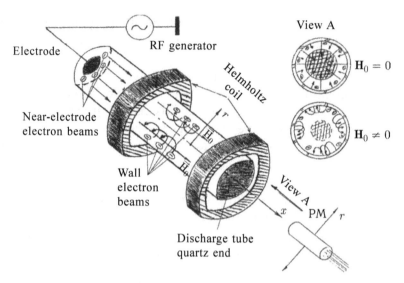

Fig. 4.7. Experimental scheme for observing wall electron beams in RFCD.

light guide, moved along the diameter of the end of the tube by an electric motor, was fed to the photomultiplier FEU-79. From the output of the photomultiplier, the signal entered the input of a selective microvoltmeter V6-10, which separated the signal at frequency f and served as a resonant amplifier, and then – to the two-coordinate Endim recorder.

For the additional possibility of monitoring WEBs and measuring the velocity of their constituent particles, a longitudinal homogeneous constant magnetic field varying in the range $0 \leq H_0 \leq 200$ Oe was applied to the RFCD using Helmholtz coils.

When the magnetic field H_0 was switched on, the near-wall luminescence was pressed against the walls, forming a thin luminous layer in the thickness of which it was possible to determine the Larmor radius of the trajectory of the beam electrons and, correspondingly, their velocity v_{ewb}. After the inclusion of the field H_0 between the near-wall luminous layer and the axial plasma column, a distinctly pronounced dark area appeared, and the glow of the plasma column decreased.

The latter was due to a decrease in the intensity of the gas excitation processes in the central region due to the deviation of the WEBs to the walls of the discharge tube.

Simultaneously with the radial distributions $I_f(r)$, similar distributions of the amplitude of the luminescence oscillations at the frequency of the second harmonic of the RF field $I_{2f}(r)$ were also obtained for obtaining additional information in the analysis of the phenomena under study.

A study of the phenomenon of the occurrence of WEBs was carried out at the RFCD in Xe at a frequency of 1.2 MHz in the pressure range $1 \leq p \leq 40$ Torr.

A typical picture of the spatial distributions $I_f(r)$ and $I_{2f}(r)$, as well as the change in these distributions when a longitudinal magnetic field is applied are shown in Fig. 4.8.

The dependence of the distributions $I(r)$, $I_{2f}(r)$ on the diameter of the outer planar electrode for the same discharge tube is seen in Fig. 4.9.

A change in the shape of the $I_f(r)$ and $I_{2f}(r)$ curves with a monotonic increase in the RF voltage $V_\sim$ supporting the discharge is shown in Fig. 4.10.

The results of a study of the dependence of the amplitude of the wall maximum in the $I_f(r)$ distribution on the voltage $V_\sim$ are shown in Fig. 4.11, while the dependence of the analogous maximum on the pressure p is shown in Fig. 4.12.

As can be seen from the experimental data, in the distributions $I_f(r)$ the central maximum due to intense high-energy near-electrode

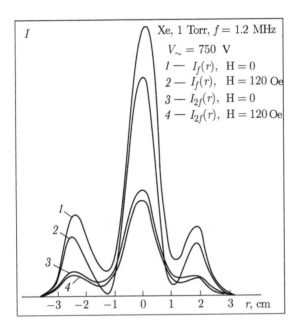

Fig. 4.8. Distribution $I_f(r)$ and $I_{2f}(r)$ with a longitudinal magnetic field and without it in the RFCD.

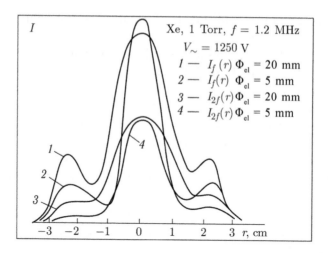

Fig. 4.9. The dependence of the distributions of $I_f(r)$ and $I_{2f}(r)$ on the RFCD on the diameter of the electrode.

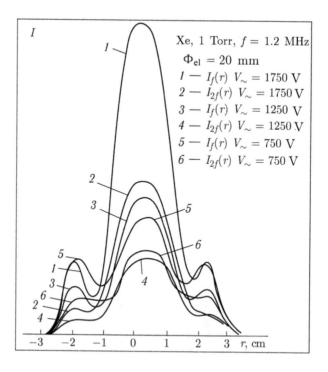

Fig. 4.10. Dependence of the distributions $I_f(r)$ and $I_{2f}(r)$ on the amplitude of the RF voltage.

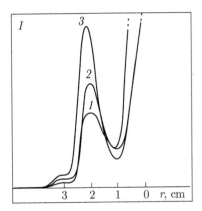

Fig. 4.11. Dependence of the amplitude of the near-wall maximum of the luminescence intensity on the RF voltage value, Xe, p = 1 Torr, f = 1.2 MHz. 1 – $V_\sim$ = 2.0 kV; 2 – 2.25 kV; 3 – 2.5 kV.

beams is accompanied by secondary maxima near the walls commensurate with the amplitude with the central one.

Thus, the WEB was experimentally observed.

The application of the magnetic field to the RFCD reduces the amplitude of the near-wall maxima of the $I_f(r)$ distributions and increases the dip between the central and wall maxima (Fig. 4.8). This means the return of the WEB by the magnetic field to the wall and indicates their radial direction from the tube wall to its axis. A decrease in the amplitude of the central maximum of the luminescence due to the absence of WEB is also noted.

The observed dependence of the $I_f(r)$ curves on the geometry of the electrodes, as expected, speaks of the dependence of the nature of the edge effects on the RFCD on the geometric factor of the discharge gap (Fig. 4.9).

The observed secondary wall maxima in the distribution $I_{2f}(r)$ can be explained by a local increase in the electron concentration at the walls, due to additional ionization produced by the WEB (Figs. 4.8, 4.10).

We also note an interesting regularity in the dependence of the $I_f(r)$ distributions on the amplitude of the RF voltage: with an increase in $V_\sim$, the relative amplitude of the wall maxima, relative to the amplitude of the central maximum, decreases substantially (Fig. 4.10). In the light of the previously expressed considerations regarding the I–V characteristic curves, the latter observation can be explained as a manifestation of the tendency of the RFCD transition

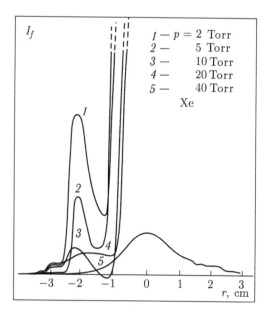

Fig. 4.12. Dependence of the wall maximum of the $I_f(r)$ distribution on the gas pressure p in the RFCD in Xe.

to a true two-electrode discharge with the closure of the discharge current on both electrodes and the reduction of the discharge of the discharge current to the surrounding 'ground'.

The behaviour of the near-wall maximum as a function of $V_\sim$ shows that with increasing $V_\sim$, the amplitude of this maximum increases, but the gap between the wall and central maxima increases (Fig. 4.1). This is explained in the first case by an increase in the plasma charge density n_e, and in the second by a certain decrease in the quasi-stationary voltage $V_\sim$ in the space charge layer near the wall as a result of the decrease in the radial component of the RF field of the edge effect, since the discharge current closes to a grounded electrode with increasing $V_\sim$. Thus, in the latter case, the energies of the WEB and the depth of their radial penetration into the depth of the discharge must decrease.

As can be seen from Fig. 4.12, as the gas pressure increases, the amplitude of the near-wall maximum of the $I_f(r)$ curves decreases, i.e., the WEB decay faster, as expected.

4.2. Methods for diagnosing electron beams in the RFCD

The above material shows that the technique of qualitative

observation of electron beams in RFCD is developed to a sufficient degree, contains a number of methods and allows obtaining sufficient information on this issue.

It is very important now to develop methods for measuring the parameters of electron beams with simultaneous determination of the characteristics of the plasma in which they propagate, which is necessary for studying the effects of beam–plasma interaction.

We note that in the literature there are as yet unknown works by other authors devoted to the experimental determination of the parameters of electron beams in RFCD.

Obviously, the main instantaneous parameters of the electron beam are: 1) the average energy of its particles $\overline{\varepsilon}_{eb}$ and 2) the density of its constituent electrons n_{eb}, where $\overline{\varepsilon}_{eb}$ is the averaged over the energy distribution function of the beam particles.

Due to the specifics of the mechanism of beam formation in RFCD, the values of $\overline{\varepsilon}_{eb}$ and n_{eb} will be significantly modulated in time over the period of the RF field, acquiring maximum values at the instants located in the vicinity of the middle of the negative half-cycle of the RF field applied to the spatial region of beam formation.

Accordingly, the values of $\overline{\varepsilon}_{eb}$ and n_{eb} obtained by different experimental methods have not the same physical meaning. Thus, using the energy analyzer method, it was possible to obtain the values of the electron beam parameters, both quasi-stationary and time-resolved. The used spectroscopic method yielded only averaged over the period of the RF field density n_{eb}.

Let us consider the diagnostic methods used in the present work to measure the parameters of electron beams.

The simplest experimental means were used to measure the order of magnitude of the time-averaged velocities of the WEB $\overline{v}_{eb}$.

The latter quantity was found from the formula for the electron Larmor radius ρ_e.

After switching on the longitudinal field $\mathbf{H}_0$, a thin, uniformly luminous layer appeared immediately near the wall of the discharge tube, after which a dark space was observed in the radial direction in front of the plasma column in the near-axis regions of the discharge (Fig. 4.8). The glow pattern was photographed through the end window of the tube and processed with an optical densitometer.

In this case, the thickness of the near-wall luminescent layer was taken as the value of $\overline{\rho}_e$.

Measurements of the average velocity of the near-electrode electron beams $\bar{v}_{eb}$ in low-pressure RFCDs were performed by the optical–magnetic method in [245].

At the same time, radiation from a small off-axis volume of the discharge was projected onto the input slit of the optical monochromator MDR-2 by the appropriate diaphragm. The magnetic field **H** applied to the discharge was oriented in the direction of observation, perpendicular to the axis of the discharge tube. This field displaced in the plane perpendicular to the observation axis, the luminous region created by the beams. By varying the intensity **H**, it was possible to project on the entrance slit of the monochromator a glow from the discharge regions caused by electron beams with different average velocities $\bar{v}_{eb}$.

The emission intensity of the extracted spectral line was transformed into a photocurrent of the photomultiplier, the signal from which was amplified and entered the input of the F-36 measuring and computing complex described in [242]. The F-36 unit produced a ramped stepped voltage, which controls the current through Helmholtz's coils. Thus, the switching of memory channels of the F-36 occurred synchronously with the change in the magnetic field. This made it possible to obtain the dependence of the intensity of the spectral lines on the field strength. The maximum of this dependence corresponds to the field strength **H**, which deflects the beam at the selected point of volume.

The average velocity of directional fast electrons was obtained using the formula for the electron Larmor radius

$$\rho_e = \frac{mc}{eH}\bar{v}_{eb}$$

from the known intensity **H** and the coordinates of the selected observation point (x_0 is measured from the tube axis, y_0 is from the electrode). From this we obtain the expression for calculating the required velocity:

$$\bar{v}_{eb} = 1.76 \cdot 10^7 \, H \frac{(x_0^2 + y_0^2)}{2y_0} \, \text{cm}/s \quad (H - \text{in oersteds})$$

This technique is characterized by high sensitivity and accuracy of measurements.

Active electrode   Active electrode   Active electrode
        c   *b*   *a*

Grounded   Grounded
electrode   electrode

Fig. 4.13. The type of discharge tubes used in the experiments: a) with symmetric electrodes, b) with asymmetric electrodes, c) with a luminophore-coated screen.

In our laboratory, we also used a simplified version of the optoelectromagnetic method for measuring $\bar{v}_{eb}$.

In this case, a discharge tube depicted in Fig. 4.13 *c* was used. The tube was a glass cylinder 18 cm in diameter, at one end of which an internal flat electrode 0.5 cm in diameter was mounted. The discharge was single-electrode. The opposite end wall of the tube was coated with a luminophore to observe the displacements of the electron beam. The distance from the electrode to the end of the tube was $y_0 = 6$ cm.

To the discharge gap perpendicular to the axis of the tube, a constant uniform magnetic field $\mathbf{H}_0$, oriented along the direction of observation, was applied and shifted the emission due to the electron beam in a plane perpendicular to the observation axis.

The centre of the spot of luminescence was fixed on the wall of the tube coated with a luminophore. A magnetic field $\mathbf{H}_0$ of 0–25 Oe was applied, the position of the centre of the glow spot was fixed, and the distance between the centres of the initial spot ($H = 0$) and the deflected one was measured.

The average velocity of the electrons of the beam $\bar{v}_{eb}$ was determined using the expression given above.

Because of the lower sensitivity of this modification of the optic-magnetic method, it was used at higher electron beam energies ($\bar{\varepsilon}_{eb} \geq 2$ keV).

As is known, information about the velocity of the beam electrons can also be obtained with the help of an energy analyzer of charged particles.

In our work we used an energy analyzer with a retarding electrostatic field [246], whose construction is described below.

It is only noted here that the discharge tubes used in experiments with RFCDs, as a rule, have flat electrodes a few centimeters in diameter. The electron beam densities in the discharge are very small. Therefore, the energy analyzer is able to detect only fluxes of the main mass of low-energy plasma electrons, and to fix electron beams its sensitivity is not enough.

True, the case of using a rather complex energy analyzer of increased sensitivity for studying the EES of the RFCD plasma is known, which was achieved by installing a secondary electronic multiplier at the analyzer output [89].

In the described studies, using the traditional energy analyzer, the sensitivity problem of this device was solved in a fundamentally different way, which consisted in increasing the beam density n_{eb} by orders of magnitude by using a sharply asymmetric RFCD with a small active electrode area and a large area of the grounded electrode. At the same time, the problem of an almost twofold increase in the energy of the beam electrons at a fixed amplitude of the applied RF voltage was simultaneously solved, which is predicted in the theoretical work [78].

As a method for measuring the density of near-electrode electron beams in a low-pressure RFCD ($p \leq 0.5$ Torr), an original spectroscopic method was used to measure the concentration of electrons emitted by the electrode n_{e0}, described in detail in chapter 2.

The possibility of identifying under these conditions the quantities n_{e0} and n_{eb} is due to the fact that, as it is easy to show, the emitted electrons pass through the NESCL practically in a collisionless regime.

Measurements of n_{e0} are made at distances from the electrode, where the ionization of the gas by the electrons of the beam has not yet begun.

Thus, at the outer edge of the NESCL, the emitted electrons gain maximum energy, the formation of the electron beam ends here.

The essence of the developed contactless technique for measuring the density of near-electrode electron beams n_{eb} is described in chapter 2.

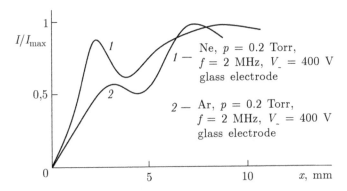

Ne, $p = 0.2$ Torr,
$f = 2$ MHz, $V_- = 400$ V
glass electrode

Ar, $p = 0.2$ Torr,
$f = 2$ MHz, $V_- = 400$ V
glass electrode

Fig. 4.14. The intensity distribution of the integral luminescence $I(x)$ in the near-electrode region of RFCD for gases Ne and Ar.

4.3. Parameters of electron beams of low-pressure RFCD

The parameters of the electron beams arising in the discharge gap depend on all the characteristics of the RFCD.

First, let us consider some experimental facts, qualitatively confirming the above statement, and then we give quantitative data of the measurements carried out.

It was noted earlier that in the luminescence distribution of $I(x)$ in the γ-type RFCD, the maximum of the glow from the electrode is the analog of the 'cathode luminescence' of DCGD (Fig. 1.33). The latter is characterized by the fact that it is excited by the flux of electrons emitted by the cathode with a density n_{e0}, which is proportional to the intensity of the luminescence. Thus, the above-mentioned emission region can be considered as an indicator of the intensity of emission processes at the electrode.

Figure 4.14 shows the distribution of $I(x)$ for RFCD with external electrodes in different gases – Ne and Ar, i.e. eere the electrodes were the glass end surfaces of the tube. It is seen that the relative amplitude of the first maximum of the distribution $I(x)$ for the pair 'Ar–glass' is much smaller than for the combination 'Ne–glass'. This should mean that in the first case the intensity of the emission processes is substantially lower than in the second.

This observation confirms the result found in the experimental work [247] that the combination 'inert gas–glass' with the lowest emissivity is due to the combination 'Ar–glass'.

The dependence of the radial distribution of the amplitude of the luminescence oscillations at the frequency of the RF field $I_f(r)$, which determines the intensity of excitation of the gas by an electron

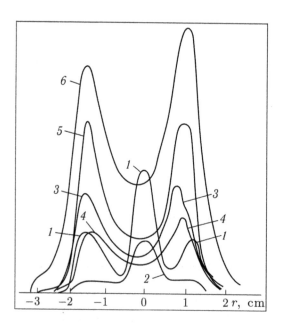

Fig. 4.15. Dependence of the distributions of $I_f(x)$ and $I_{2f}(x)$ on the geometry of the electrodes of RFCD, Xe, $p = 1$ Torr, $f = 1$ MHz, $V_{\sim} = 500$ V. $1 - I_f(r)$, $D_{eл} = 8$ mm; $2 - I_{2f}(r)$, $D_{eeл} = 8$ mm; $3 - I_f(r)$, el.-ring $D_{eл} = 64$ mm; $4 - I_{2f}(r)$, el.-ring $D_{eл} = 64$ mm; $5 - I_f(r)$, el.-'cup', $h = 10$ mm; $6 - I_{2f}(r)$, el.-'glass", $h = 10$ mm.

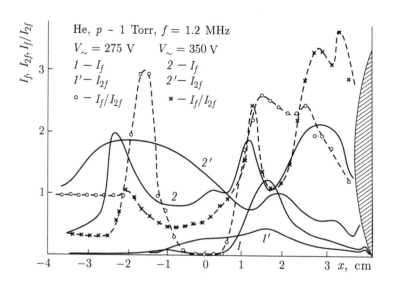

Fig. 4.16. The spatial distribution of the amplitudes of the luminescence oscillations at the field frequencies f and $2f$ and the ratio of these amplitudes I_f / I_{2f} in the RFCD with a convex active electrode.

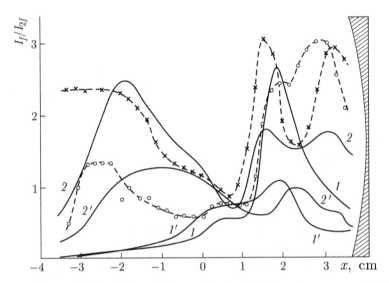

Fig. 4.17. The spatial distribution of the amplitudes of the luminescence oscillations $I_f(x)$ and $I_{2f}(x)$ in the RFCD with a concave active electrode He, $p = 1$ Torp, $f = 1.2$ MHz, $V_\sim = 275$ V. $1 - I_f$, $1' - I_{2f}$, $V_\sim = 350$ V; $2 - I_f$, $2' - I_{2f}$.

beam, on the geometry of the outer electrodes (a flat electrode with a diameter of 8 mm, an electrode with a 'cup' end with a height $h = 10$ mm, an annular electrode with a diameter $D_{el} = 64$ mm) is shown in Fig. 4.15. Significantly different distributions of $I_f(r)$ for the two reduced configurations of the electrodes indicate differences in the parameters of electron beams and their spatial distributions in these cases.

The possibility of focusing electron beams in the discharge gap of the RFCD was studied experimentally, using electrodes of both a convex and concave shape with a radius of curvature of $R_{el} = 4.5$ cm.

The longitudinal distributions of $I_f(x)$ in RFCD with an active electrode of a convex shape are shown in Fig. 4.16, and with an electrode of a concave shape – in Fig. 4.17. As can be seen, the distributions of $I_f(x)$ for electrodes of different signs of curvature are significantly different.

The spatial variation of the $I_f(x)$ curves for an electrode with a concave surface reveals regions of a significant increase, in comparison with analogous curves for a convex electrode, of the amplitude of the signal at the frequency of the RF field (Fig. 4.17), which can be interpreted as a local increase in the electron beam density due to the effect of its focusing.

A very clear change in the behaviour of electron beams in RFCD was manifested when comparing the luminescence pattern of this discharge in the V_0-mode and in the I_0-mode (Fig. 4.18). If in the first case only an intense beam from the active electrode was observed, then in the second case two opposing beams were observed from both electrodes, and a beam of a large cross section was observed from a large-area grounded electrode, respectively.

It was noted above that a decrease in the area of the active electrode made it possible to increase by one or two orders of magnitude the beam density n_{eb}, which made it possible to study the beam parameters using a standard energy analyzer of charged particles.

We present the quantitative data obtained for measurements of the parameters of electron beams in the RFCD.

Typical results of measurements by a magneto-optical method in a discharge tube with a luminescent screen of the average electron energies of the beam $\bar{\varepsilon}_{eb}$ as a function of the amplitude of the RF voltage are shown in Fig. 4.19. These measurements were made in a single-electrode RFCD with a small electrode surface area (diameter 0.5 cm). Since the discharge current in this case was closed to the 'ground' through the space surrounding the discharge gap, the area of the second 'electrode' was much larger than that used, and a sharply asymmetric discharge was investigated.

Let us compare these experimental results with the theoretical prediction [78], according to which for our conditions the maximum energy of the beam electrons should approach the value $\bar{\varepsilon}_{eb} \approx 2$ eV$_{\sim}$. As can be seen from the graph of Fig. 4.19, the experimental and theoretical values of $\bar{\varepsilon}_{eb}$ are very close.

The results of measurements of the time-averaged velocities of the near-electrode electron beams $\bar{v}_{eb}$ obtained by the optical-magnetic

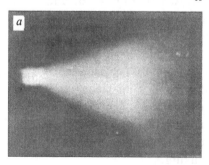 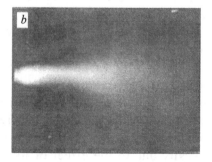

Fig. 4.18. Type of glow discharge in I_0- (*a*) and V_0-modes (*b*) (air, $p = 0.08$ Torr, $f = 1$ MHz, $V = 750$ V, $D_a = 8$ mm, $D_g = 56$ mm).

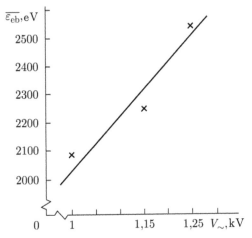

Fig. 4.19. Dependence of the average energy of electron beams on the amplitude of RF voltage of RFCD in air, $p = 0.1$ Torr, $f = 1.2$ MHz.

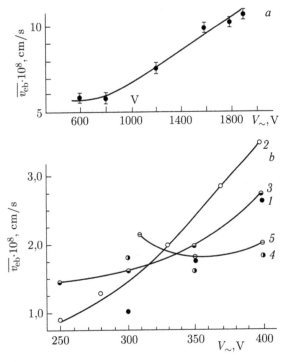

Fig. 4.20. Dependence of the averaged velocity of electron beams v_{eb} on the amplitude of the RF field voltage $V_{\sim}$ RFCD in He. $a - f = 1$ MHz, $p = 0.2$ Torr; $b - f = 100$ MHz, $1 - p = 0.22$ Torr, $2 - 0.10$ Torr, $3 - 0.08$ Torr, $4 - 0.06$ Torr, $5 - 0.14$ Torr.

spectroscopic method of the Leningrad State University are presented in Fig. 4.20 [245].

In these two experiments, the dependence of $\bar{v}_{eb}$ on the amplitude of RF voltages for very different frequencies was studied – 1 MHz, 100 MHz.

At a low frequency, only one real near-electrode mechanism for generating accelerated electrons is possible: the acceleration of electrons emitted by the electrode in the NESCL. In this case, similar to the above results obtained by the less sensitive magneto-optical method (Fig. 4.19), monotonous growth of $\bar{v}_{eb}$ is also observed with increasing $V_\sim$ (Fig. 4.20 *a*).

At a frequency of 100 MHz, according to the estimates [248], with pressures $p < 0.1$ Torr, a second effective mechanism appears- stochastic acceleration of 'plasma' electrons by an oscillating potential barrier of the NESCL [56]. Thus, a mechanism competing with the first additional discharge maintenance mechanism appears.

Analysis of the curves $\bar{v}_{eb}(V_\sim)$ in Fig. 4.20 shows that at a frequency of 100 MHz in the voltage range $V_\sim = 250-400$ V, the electron beam energies are in a narrow interval $\bar{\varepsilon}_{eb} = 3-35$ eV. The orders of the valuees of $\bar{\varepsilon}_{eb}$ found under these conditions show that they are characteristic of the mechanism of stochastic acceleration of electrons.

Note that the mechanism mentioned above is very sensitive to the gas pressure. With $p \geq 0.15$ Torr (curves *1*, *2*) the velocities $\bar{v}_{eb}$ depend strongly on $V_\sim$, which is not typical for the pure mechanism of stochastic acceleration. Thus, in this case one should also assume the presence of a weak mechanism for the acceleration of γ-electrons in the NESLC, along with the first mechanism.

At $p = 0.1$ Torr curve 3 shows an increasing transition to the mechanism of electron acceleration by the oscillating boundary of the NESCL. And even at $p < 0.08$ Torr (curves 4, 5) there is practically no $\bar{v}_{eb}(V_\sim)$ dependence, which indicates the prevailing character of the stochastic electron heating mechanism in maintaining RFCD.

Concluding our discussion of the characteristic energies of electron beams in the discharge under study, we note that in one of the experiments with RFCD in Xe at $p = 0.08$ Torr, $V_\sim = 750$ V, $f = 1.2$ MHz and $H = 120$ Oe, the electron Larmor radius was about 0.5 cm. It was established from this that the average energy of the WEB was $\bar{\varepsilon}_{eb} \approx 280$ eV.

The last physical value shows that the wall electron beam (WEB) can return to the RFCD an appreciable fraction of the energy of the RF field leaving as a result of edge effects.

Taking into account the justified identification of the concentration of electrons emitted from the electrode surface n_{e0} with the density of the near-electrode electron beams n_{eb}, we can consider all the experimental data about the density of the emitted electrons n_{e0} as the results of studying the density of electron beams n_{eb} as a function of the RFCD parameters.

Thus, on the basis of the experimental data mentioned above, it can be asserted that in this paper we present working methods for measuring n_{eb} in RFCD, both with internal and external electrodes; It was established that under similar physical conditions the values of n_{eb} in the cathode region of the TPGT and the near-electrode region of the RFCD are close in magnitude; The density of near-electrode n_{eb} beams in RFCD increases with increasing RF voltage and field frequency.

The results of the present study suggest that there are real possibilities for a substantial increase in the values of the parameters of the electron beams n_{eb} and $\bar{\varepsilon}_{eb}$ in RFCD and the conversion of the latter into a beam-plasma discharge maintained both directly by the beam itself and by beam-plasma effects initiated by it.

The results of the research below confirm this statement.

In order to significantly increase the values of the parameters of the near-electrode electron beams in the RFCD, the following recommendations can be made: a) to increase the beam electron energy $\bar{\varepsilon}_{eb}$, one should use the voltages $V_{\sim} \gg 1$ kV and an asymmetric discharge with a small area of the active electrode; b) an increase in the density n_{eb} beams can be obtained by: 1) decreasing the area of the active electrode; 2) using the voltages $V_{\sim} \gg 1$ kV, at which the energy of the ion bombarding ions $\varepsilon_i \gg 1$ keV, which leads to a transition from the potential emission of electrons to the kinetic energy with an emission coefficient $\gamma > 1$; 3) using higher frequencies of the field $f \gg 1$ MHz; 4) optimally choosing the combination 'working gas–electrode material'; 5) by varying the geometry of the electrodes, allowing, for example, focusing of electron beams or the effect of the RF effect – 'hollow cathode'.

4.4. The processes of spatial and temporal relaxation of near-electrode electron beams

In order to determine the characteristic spatial scale of the influence of the boundary effects forming the near-electrode plasma, we have experimentally studied the processes of relaxation of electron beams with respect to momentum and energy.

This is of considerable interest, since the near-electrode beams are the most important factor in the formation of the electron energy spectrum (EES) of the near-electrode plasma.

As is known, the spatial distribution of the intensity of the spectral lines $I_\lambda(x)$ reflects the relaxation of beams with respect to the energy εeb.

The spatial course of the relaxation process of the near-electrode beam with respect to the momentum is characterized by the distribution of the degree of polarization of the radiation of the spectral lines $P_\lambda(x)$, when the appearance of the emission spectrum and its polarization are produced in the process of direct excitation of atoms by fast beam electrons [216].

We propose a second, independent method for controlling the process of beam relaxation with respect to momentum by recording the spatial distribution of the oscillation amplitude of the intensity of the spectral lines or the integral luminescence at the frequency of the RF-supporting field $I_{\lambda f}(x)$, $I_f(x)$.

This method is based on the experimentally established fact of pulsation of near-electrode beams with a field frequency f [51, 67]. It has been noted above that the presence of oscillations of the glow at a frequency f at any point in the discharge gap indicates the penetration of a near-electrode electron beam into the given place.

This polarization method is highly sensitive and very informative, but it requires rather complicated equipment and, unfortunately, is only suitable at rather low pressures ($p \leq 10^{-2}$ Torr), when collisions of electrons with excited atoms do not produce a strong depolarizing effect [216].

The advantage of the proposed new method is that it has no limitations on the gas pressure. It should also be noted that, generally speaking, this method is applicable not only in conditions of a RF discharge with natural modulation of beam parameters.

It can be used also in any stationary conditions, if the investigated near-electrode beam is modulated in time.

Due to the difference in the possibilities of the diagnostic methods and the physical conditions in the discharge, two series of investigations of the relaxation of electron beams at elevated ($10^{-1} \ll p \leq 10$ Torr) and lower ($10^{-3} \leq p \leq 10^{-1}$ Torr) pressures were carried out.

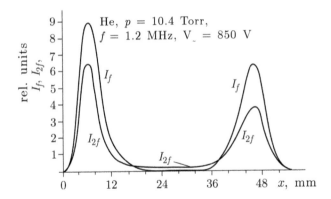

Fig. 4.21. The spatial distribution of the amplitude of the luminescence oscillations $I_f(x)$ and $I_{2f}(x)$ in the RFCD.

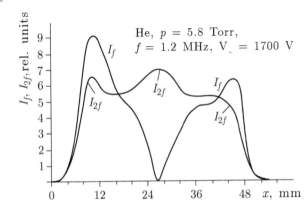

Fig. 4.22. The spatial distribution of the amplitude of the luminescence oscillations $I_f(x)$ and $I_{2f}(x)$ in the RFCD.

Let us first consider the results of studying the behaviour of the near-electrode beams in the first of the indicated pressure ranges, obtained by analyzing the distributions of $I_f(x)$.

The characteristic distributions of $I_f(x)$ (and also some distributions of the luminescence oscillations at the frequency of the second harmonic of the RF field $I_{2f}(x)$) are shown in Figs. 4.21–4.24.

It should be noted that the correct application of this technique assumes that there is no spatial coverage of the beams from the opposing electrodes in the discharge, which was quite possible under the appropriate conditions for maintaining RFCD.

An example of the distribution of $I_f(x)$ with non-overlapping beams is shown in Fig. 4.21, and in case of overlapping – in Fig.

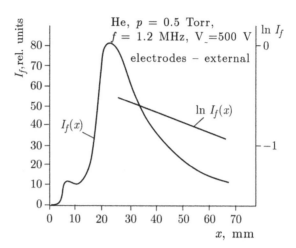

Fig. 4.23. The spatial distribution of the amplitude of the luminescence oscillations $I_f(x)$ in the RFCD.

4.22. One can distinguish between these two physical situations by the sign of the curvature of the distribution of $I_f(x)$ in the direction of the centre of the discharge: in the first case, the sign of the curvature is negative (Fig. 4.21), and in the second – positive (Fig. 4.22). In the latter case, the beams do not abruptly decay completely in the centre, but simply when two signals at the frequency f in the antiphase are added at the centre, they are perceived as a second-harmonic signal. This is clearly illustrated by the spatial distribution of $I_{2f}(x)$ in Fig. 4.22.

It is obvious that the distribution section of $I_f(x)$ from the electrode to the maximum of this curve corresponds to the formation space of the electron beam. The course of the $I_f(x)$ curve from the maximum toward the centre of the discharge gap demonstrates the decay of the beam in the plasma.

The construction of the $\ln I_f(x)$ curves in the decay regions of the beams showed their rectilinear nature, that is, the exponential decay law of the beams with the characteristic relaxation length of the beam with respect to the momentum l_p (Fig. 4.23).

Thus, in the damping region of the beam, the curve $I_f(x)$ can be present in the form:

$$I_f(x) = I_{f0} \exp\left(-\frac{x}{l_p}\right),$$

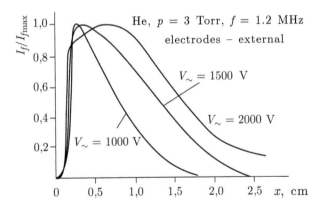

Fig. 4.24. The spatial distribution of the ratio $I_f(x) / I_{f\max}(x)$ for different amplitudes of the RF voltage $V_\sim$.

where I_{f_0} is the maximum amplitude of the luminescence oscillations at the frequency f in relative units, and the coordinate x is measured from the point of finding the maximum amplitude I_{f_0}.

According to the data of the present work, the near-electrode electron beams in the RFCD were sufficiently monoenergetic. Therefore, we assume that the spatial damping of the beam, reflected by the curve $I_f(x)$, means a decrease in its density $n_{eb}(x)$ and can be represented as:

$$n_{eb}(x) = n_{e0} \exp\left(-\frac{x}{l_p}\right),$$

where n_{e0} is the maximum density of the beam electrons at point $x = 0$.

From this we can determine the value l_p using the equation

$$\ln\left[\frac{n_{eb}(x)}{n_{e0}}\right] = -\frac{x}{l_p} = -\kappa x,$$

where $1/l_p = \kappa = \operatorname{tg} \alpha$ is the tangent of the angle of inclination of the given straight line to the axis x.

The straight line $\ln\left[\dfrac{n_{eb}(x)}{n_{e0}}\right] = -\kappa x$ constructed by this procedure is shown in in Fig. 4.23 where the slope of the straight line is characterised by the value tg α = 0.46. Correspondingly, we obtain the relaxation length of the beam.

$$l_p = \frac{1}{tg\alpha} = \frac{1}{0.46} = 2.16 \text{ cm.}$$

To determine the mechanism of beam attenuation, from the experimental data, we determine the characteristic cross section for the collisions of the beam electrons with the atoms $\sigma_{en}(v)$.
As is known [157],

$$l_p = \frac{1}{p_0 P_c} = \frac{1}{p_0 n_a \sigma_{en}(v)} \quad \text{and} \quad \sigma_{en}(v) = \frac{1}{l_p p_0 n_a},$$

where p_0 is the reduced pressure, P_c is the probability of collisions of an electron with atoms at a length of 1 cm for p = 1 Torp, n_a is the concentration of atoms at p = 1 Torr.
For the conditions of the experiment under consideration (Fig. 4.23) we have:
He, p = 0.5 Torr, n_a = 3.54 · 10^{16} cm^{-3}, l_p = 2.16 cm,

$$\sigma_{en}(v) = \frac{1}{2.16 \cdot 0.5 \cdot 3.54 \cdot 10^{16}} = 0.3 \cdot 10^{-16} \text{ cm}^2.$$

Thus, the length l_p corresponds to the beam-scattering mechanism for paired elastic collisions with atoms with the usual gas-kinetic cross section $\sigma_{en} \sim 10^{-16}$ cm^2 [4].
The experimental results in Fig. 4.24 show an increase in l_p with increasing RF voltage:

$$V_\sim = 1000 - 2000 \text{ V}, \quad l_p = 0.7 - 1.3 \text{ cm.}$$

In accordance with the mechanism of relaxation of electron beams under these conditions, the characteristic relaxation time of the beam with respect to the momentum is $\tau_p \sim 1/v_{bn}$, where v_{bn} is the collision frequency of the beam electrons with atoms. We have:

$$v_{bn} = n_a \sigma_{en} v_{eb}, \quad n_a \sim 10^{16} \text{ cm}^{-3}, \quad \sigma_{en} \sim 10^{-16} \text{ cm}^2, \quad v_{eb} \sim 10^9 \text{ cm/s,}$$

$$v_{bn} \sim 10^{16} \cdot 10^{-16} \cdot 10^{9} = 10^{9}\ \text{s}^{-1}, \quad \text{from which} \quad \tau_{p} \sim \frac{1}{v_{bn}} \sim \frac{1}{10^{9}} \sim 10^{-9}\ \text{s}.$$

Let us now consider the experimental results of studying the relaxation of electron beams at lower pressures ($p \ll 10^{-1}$ Torr) obtained using a magnetic polarization spectrometer described in [216].

The characteristic spatial distributions of the degree of polarization of the radiation of the spectral lines $P_{\lambda}(x)$ in the near-electrode regions of RFCD in Ar and He are shown in Figs. 2.33, 2.34. Simultaneously, the radiation intensity distribution of the same spectral lines $I_{\lambda}(x)$, also shown in the above-mentioned figures, was studied.

Taking into account that the distributions $I_{\lambda}(x)$ and $P_{\lambda}(x)$ reflect the spatial relaxation of the beams in energy and momentum, respectively, we note from the observed distributions the faster pulse relaxation with respect to momentum than in energy.

Processing exponentially decreasing distributions $P_{\lambda}(x)$ in the RFCD in Ar in the above way (for the $I_{\lambda}(x)$ distributions), we determine the relaxation length with respect to the momentum lp along the slope of the curve $\ln (P_{\lambda}(x)/P_{\lambda\,\text{max}}) = -x/l_{p} = -x\ \text{tg}\ \alpha$ (Fig. 4.23).

The following values of l_{p} for different RF voltages are obtained:

$$V_{\sim} = 210\ \text{V}, \quad l_{p} = 0.58\ \text{cm}, \quad \lambda_{eb} = 2.8\ \text{cm},$$
$$V_{\sim} = 180\ \text{V}, \quad l_{p} = 0.52\ \text{cm}, \quad \lambda_{eb} = 2.6\ \text{cm},$$
$$V_{\sim} = 85\ \text{V}, \quad l_{p} = 0.42\ \text{cm}, \quad \lambda_{eb} = 2.3\ \text{cm},$$

where the mean free paths of the beam electrons along elastic collisions with atoms λ_{eb} are also given.

The observed rapid decay rate of the $P_{\lambda}(x)$ curve in the vicinity of the electrode can not be explained by electron–atom collisions, since the characteristic spatial scale of this decay (the length l_{p}) is several times smaller than the mean free path of the beam electrons λ_{eb}.

Apparently, in this case, the cause of the anomalously fast relaxation of electron beams with respect to the momentum is the appearance of a beam–plasma instability in the discharge, on the intense oscillations of the electric potential of which the beam is effectively scattered.

The beam–plasma instabilities in the RFCD under study are discussed in more detail below. Here we confine ourselves to quantitative estimates of the possible parameters of such an instability and the characteristic relaxation lengths of the beam l_p due to it.

Binding the decay of the distribution $P_\lambda(x)$ to the Cherenkov damping mechanism of the electron beam, we estimate the growth rate of longitudinal waves δ [155]

$$\delta \approx \omega_{e0}\left(\frac{n_{eb}}{n_e}\right)^{1/3} = 5.7\cdot10^4\left(n_{eb}\,n_e^{1/2}\right)^{1/3}\ \mathrm{s}^{-1}.$$

We determine the order of the value $n_{eb} = \dfrac{(1/4)n_i v_i \gamma}{u}$, where $n_i \approx n_e$, $v_i = \left(\dfrac{kT_e}{M_i}\right)^{1/2}$ is the velocity of the ions at the boundary, $u = \left(\dfrac{2eV_\sim}{m}\right)^{1/2}$ is the velocity of the beam electrons in the order of magnitude; γ is the coefficient of ion–electron emission.

Taking into account that the lifetime of the electron beam is $\tau \sim T/2 = 1/2f$, we obtain the condition necessary for the development of the beam instability:

$$\delta > v_{en},\, 2f.$$

We make estimates for the experimental conditions:

He, $p = 2\cdot10^{-2}$ Torr, $V_\sim = 100$ V, $f = 10^8$ Hz, $n_e = 2\cdot10^{10}$ cm^{-3}, $T_e \sim 5\cdot10^4$ K, $v_{en} = 5\cdot10^7$ s^{-1}, $\gamma \sim 10^{-1}$.

Since in this case $2f > v_{en}$, the condition for the development of the instability is:

$$\delta > 2f.$$

Taking into account the above, it is easy to obtain

$$\delta = 6.2\cdot10^2 n_e^{1/2} T_e^{1/6}\gamma^{1/3}V_\sim^{-1/6}$$

and further, substituting the numerical values of the quantities, we have:

$$\delta = 4.4 \cdot 10^8 \text{ s}^{-1} > 2f = 2 \cdot 10^8 \text{ s}^{-1}.$$

From this it is clear that under the experimental conditions the threshold of excitation of beam instability is reached. In this case, the beam relaxation time with respect to the momentum should be of the order of $\tau_p \sim \delta^{-1}$, which for the considered conditions is:

$$\tau_p \sim \frac{1}{4.4 \cdot 10^8} \approx 2 \cdot 10^{-9} \text{ s.}$$

Note that to carry out the above estimates it is sufficient to know the values of n_e, T_e, γ, $V_\sim$ with a relatively low accuracy due to their corresponding power-law dependences – $n_e^{1/2}, T_e^{1/6}, \gamma^{1/3}, V_\sim^{-1/6}$.

The characteristic relaxation length of the beam l_p is estimated from the well-known formula:

$$l_p \sim \frac{u}{\delta}.$$

We have:

$$l_p \approx \frac{5.9 \cdot 10^8}{4.4 \cdot 10^8} = 1.3 \text{ cm.}$$

The processing of the experimental curve $P_\lambda(x)$ for these experimental conditions in Fig. 2.33 (curve P_1) gives the value of the relaxation length of the beam $l_p \approx 1$ cm. This value is quite close to that obtained above as a result of an estimate assuming that the beam relaxation rate is determined by the arising beam–plasma instability.

The mean free path of the beam electrons in the gas calculated for the experimental conditions was $\lambda_{eb} \approx 12.5$ cm.

Thus, it is confirmed that the relaxation of the near-electrode beams under the investigated conditions is not due to non-elastic collisions of beam electrons with atoms, but another, more efficient mechanism – beam scattering under the action of the arising beam-plasma instability.

In conclusion, a general consideration should be given. On the basis of the material presented above, it can be concluded that the near-electrode electron beams in the low-pressure RFCD are a particular manifestation of the universal physical situation: the surface of any material under the RF potential, immersed in a low-pressure gas environment, becomes the emitter of electron beams with parameters determined by the physical conditions of their occurrence.

This fundamental physical phenomenon must have a number of physical effects and numerous practical applications.

As an example, here are some of them:

1) the formation of wide, right up to the multi-keV range, EES plasma with the help of relaxing electron beams;

2) the creation by the simplest technical means of electron beams of various configurations, including beams of a large cross-sectional area, for electronically processing surfaces with dosed energies and electron flux density;

3) creation of electron guns of a multi-keV range for impact on objects inside closed dielectric cavities;

4) excitation by electron beams of luminophores in illuminating lamps.

Physical properties of RF gas discharge plasma

5.1. Features of the properties of RFCD plasma

The main goal of studying the physical mechanism of the RF discharge is to control the discharge parameters, to obtain a gas-discharge plasma with a specified density of charged particles n_e and a function of the electron energy distribution $f_e(\varepsilon)$.

We recall the definition of plasma as a special, fourth state of matter. There is no exhaustive, rigorous definition of the concept of 'plasma'. We will use such working formulation:

plasma is a partially or completely ionized gas satisfying two conditions:

1) the quasi-neutrality $n_e \approx n_i$, where n_e is the electron density, n_i is the concentration of singly charged positive ions;

2) $r_{De} \ll L$, where $r_{De} = \sqrt{\dfrac{kT_e}{4\pi e^2 n_e}}$ is the Debye charge shielding radius, L is the characteristic size of the region occupied by the ionized gas.

We note the main features of the properties of low-pressure RFCD plasma.

There is a significant dependence of the parameters and properties of the plasma on the frequency of the RF field at a fixed amplitude of the RF voltage.

Depending on the ratio of the duration of the period T of the field and the characteristic times of the diffusion escape of electrons to the walls of the discharge tube and the relaxation of the electron energy distribution function $f_e(\varepsilon)$ for a time T can be stationary or variable, both the concentration of charged particles n_e and the distribution function $f_e(\varepsilon)$. In this case, the function $f_e(\varepsilon)$ can change in time in whole or in separate energy intervals, which will be considered below.

As noted, in particular, in work [6], the plasma has the properties of both an ohmic conductor and a dielectric. In this case, in the case of low field frequencies, the plasma manifests itself more as a pure conductor, and at higher frequencies its dielectric properties become more pronounced.

The expressions for the conductivity and permittivity of the plasma in the RF field have the form:

$$\sigma = \frac{e^2 n_e v_{en}}{m(\omega^2 + v_{en}^2)},$$

$$\varepsilon = 1 - \frac{4\pi e^2 n_e}{m(\omega^2 + v_{en}^2)}.$$

The problems of plasma electrodynamics under the conditions of RFCD are briefly considered in [6]. A recently published monograph [7] has been dedicated to the physical properties of the quasi-equilibrium RF plasma.

A distinctive feature of the investigated RFCD plasma is the relative simplicity of controlling its parameters, in particular, the energy of electrons. In certain regimes with high RF voltages, anomalously high electron temperatures T_e of the order of 10^6 K were recorded [169].

The most characteristic feature of the plasma described below RFCD is its strong non-equilibrium, significant control capabilities both in the form and width of the energy range of the electron energy spectrum of the plasma.

Since the EES (electron energy spectrum) determines the nature of all the elementary processes taking place in the plasma, in studying the physical properties of the RFCD plasma, the main emphasis in the described experimental studies was made on the study of the energy spectrum of electrons and the development of a technique for its formation.

To fully take into account the features of the properties of the RFCD plasma, one should mention the possibility of creating a plasma even in a highly aggressive chemical medium by means of a discharge with external electrodes.

5.2. The nature of motion of charged particles in the RFCD

Practically, the active component of the electric current in gas discharges with an alternating field is transferred by electrons.

In a gas-discharge plasma, the electron velocity has a chaotic and directional components. In RFCD, the directional velocity component is related to the vibrational motion of electrons in the harmonic field $E = E_a \sin \omega t$, which determines a number of the physical properties of the RF discharge. The average velocity of the chaotic motion $\bar{v}_e$ of is usually large compared to the velocity of the directed, vibrational motion $|v_a|$, exceeding it to about two orders of magnitude.

The equation of motion for the electron velocity v, averaged over many elastic collisions with atoms, has the form [4]:

$$m\dot{v} = -e\mathbf{E}_a \sin \omega t - mv\, v_{en},$$

where v_{en} is the effective collision frequency of an electron with atoms associated with the momentum transfer. With the help of this equation, we find expressions for the velocity **v** of the vibrational motion and the resulting electron displacement **r**:

$$\mathbf{v} = \frac{e\mathbf{E}_a}{m\sqrt{\omega^2 + v_{en}^2}}\cos(\omega t + \varphi), \quad \varphi = \operatorname{arctg}\frac{v_{en}}{\omega},$$

$$\mathbf{r} = \frac{e\mathbf{E}_a}{m\omega\sqrt{\omega^2 + v_{en}^2}}\sin(\omega t + \varphi),$$

where φ is the phase shift between the velocity **v** and the field strength vector **E**. Hence, in the absence of collisions of an electron with atoms ($v_{en} = 0$), the velocity **v** is shifted in phase with respect to the field **E** by $\pi/2$. Therefore, the work done by the field in 1 s on the electron, averaged over the period of the field, in this case is equal to $\langle e\mathbf{E}/v \rangle = 0$.

Electrons can be heated in the RF field only in the presence of collisions. In the interval between collisions, the electron acquires vibrational energy. In the event of a collision, it arbitrarily changes its direction of motion practically without changing the velocity modulus. The energy which the electron at that moment possesses belongs to the chaotic, since it begins to accelerate anew. Therefore, each act of collision is accompanied by a transition to chaotic energy of the magnitude of the order of the energy acquired in the interval between collisions, which is close to the average vibrational energy.

The vibrational motion of an electron with an amplitude of displacement

$$A = \frac{eE_a}{m\omega\sqrt{\omega^2 + v_{en}^2}}$$

is superimposed on the chaotic motion.

In the limiting case of very frequent collisions ($v_{en}^2 \gg \omega^2$) the vibrational velocity

$$\mathbf{v} \approx -\frac{eE_a}{mv_{en}}\sin \omega t = -\frac{e}{mv_{en}}\mathbf{E}(t) = -\mu_e \mathbf{E}(t) = \mathbf{v}_{dr},$$

where $\mu_e = \dfrac{e}{mv_{en}}$ is the mobility of an electron, at each moment of time coincides with the drift velocity of electrons in a constant field equal to the instantaneous one.

Let us compare the values of the velocities of the average chaotic $\overline{v}_e$ and the drift $\mathbf{v}_{dr}$ for a particular case of a RF discharge with the following parameters:

The working gas is He, $p = 0.5$ Torr, the frequency of the field is $f = 6$ MHz, $V_\sim = 600$ V, $E_\sim = 5$ V/cm,

$$T_e = 5 \cdot 10^4 \text{ K}, \quad v_{en} \approx 10^9 \text{ s}^{-1}.$$

$$\overline{v}_e = \left(\frac{3kT_e}{m}\right)^{1/2} = \left(\frac{3 \cdot 1.38 \cdot 10^{-16} \cdot 5 \cdot 10^4}{9.1 \cdot 10^{-28}}\right)^{1/2} = 1.6 \cdot 10^8 \text{ cm/s}.$$

$$\mathbf{v}_{dr} = \frac{eE_\sim}{mv_{en}} = \frac{4.8 \cdot 10^{-10} \cdot 5}{3 \cdot 10^2 \cdot 9.1 \cdot 10^{-28} \cdot 10^9} \approx 10^7 \text{ cm/s}.$$

The obtained values of $\overline{v}_e$ and $\mathbf{v}_{dr}$ show the characteristic ratio of the components of the velocity of chaotic and directed motion of electrons in a gas-discharge plasma.

Usually in the radiofrequency range, it is believed that the ions are sufficiently inert to respond to the change in the RF field. Therefore, the motion of ions is mainly considered in quasi-stationary electric fields arising in the RF discharge.

5.3. Types of low-pressure RFCD plasma

The physical properties of the low-pressure RFCD plasma, depending on the mode of its maintenance, can vary considerably. We can conditionally distinguish several types of plasma of a given discharge:

1) an α-type RFCD,

2) two kinds of γ-type RFCD-analogues of the plasma of the regions of 'negative glow' (NG) and 'positive column' (PC) DC glow discharge,

3) the plasma of a symmetrical RFCD with a large area electrodes,

4) a plasma of sharply asymmetric RFCD (ARFCD) with a small area of the active electrode,

5) ARFCD plasma in the beam–plasma discharge regime.

The most significant difference between these types of plasma is the electronic energy spectrum (EES) inherent in each variety, which is discussed in detail below.

As is well known [6], the RF plasma of the α-discharge is supported by the RF field at each point in the volume of the discharge gap and has an EES close to the Maxwellian EES.

The physical properties of the γ-type RFCD are very close to the properties of DCGD, including the kind of spatial structure containing the plasma regions of the negative glow and the positive column with close characteristic parameters.

The main differences with DCGD are due to the presence in the RFCD of a new fundamental parameter – the frequency of the RF field.

The plasma of a symmetrical RFCD with a large surface of the electrodes is characterized by a high content of high-energy electrons at high RF voltages, however, the relative density of such electrons is rather small.

The sharply asymmetrical RFCD, due to the small area of the active electrode, is characterized by relatively high densities of near-electrode electron beams and high-energy plasma electrons, sufficient for reliable registration through a power analyzer of charged particles with a decelerating electric field.

Finally, in ARFCD, the excitation regimes of the beam–plasma instabilities were reached and there was an RF beam–plasma discharge with anomalously high temperatures of plasma electrons $T_e \leq 10^6$ K [169]. The corresponding experimental results are given below.

5.4. Electron energy spectrum of plasma of symmetrical RFCD

Traditional RFCD research, as a rule, is devoted to the study of plasma in the central region of the discharge, which is close in properties to the 'positive column' plasma of DCGD with a quasi-equilibrium energy distribution of electrons. Such a plasma has a rather narrow EES (one or tens of eV), very conservative when the external parameters of the discharge change in a wide range.

One of the main goals of writing this chapter was presentation of the results of an experimental study of the electronic energy spectrum of a near-electrode plasma created by the RFCD boundary processes, as well as the possibilities of controlling the plasma electron energy distribution.

The insufficiently studied near-electrode plasma of RFCD, in contrast to the plasma of the central regions of discharges, is very sensitive to changes in the external characteristics of the discharge.

A characteristic feature of the plasma in question is the formation of non-equilibrium EESs that extend in the range up to several keV, with a corresponding increase in the applied RF voltage.

In the presented studies, the physical conditions of the experiment varied in such a way that the mechanism of formation of the EES of the plasma significantly changed.

In connection with the important role of near-electrode electron beams in the process of formation of plasma EES, we note that the literature described experiments with RFCD with electrodes of a sufficiently large area. Therefore, the density of the near-electrode electron beam (NEEB)becomes small and in the experiments conducted a study of their role in the discharge mechanism is very difficult because of the insufficient sensitivity of the diagnostic methods used.

Taking into account the foregoing, in the described studies, along with symmetrical electrodes of planar geometry, we used ARFCDs with a small area of the active electrode.

As a result, for a fixed RF voltage $V_\sim$, the density of the near-electrode beams is increased by one or two orders of magnitude, the electron energy of the beam ε_{eb} in a sharply asymmetric discharge is increased almost twofold, and it became possible to study in detail the high-energy part of the EES of the plasma and the beam with a conventional energy analyzer with a decelerating electric field. Accordingly, in experiments with RFCDs with symmetrical electrodes, emphasis was placed on the study of the distribution of the most numerous group of 'slow' electrons of the distribution function $f_e(\varepsilon)$.

We also note that the results of investigations in a stable gas-discharge plasma are presented here.

5.4.1. A review of the literature data

Let us briefly dwell on the known literature data reflecting the current level of research on this issue.

While relatively few measurements of EES in RFCD are conducted, and, as a rule, averaged over time. The overwhelming majority of them are conducted by a contact, probe method, which has a number of problematic moments. In part, in this connection, considerable efforts have been directed to theoretical methods of studying the EES of plasma of RFCD.

So far, there is no complete theoretical method that allows one to describe in detail the RFCD, including the electronic kinetics of its plasma.

In a wide range of physical conditions, the EES of RFCD can be determined both by local conditions in space by conditions and by processes in the entire volume of the discharge, can be either quasi-stationary or vary with time.

A number of theoretical approaches to the determination of EES in RFCD are given in the literature [251, 120, 125, 256].

The basic principles of describing the spatio-temporal behavior of EES with the help of the kinetic equation are well known [256]. However, in general this equation is still not solved due to mathematical and technological difficulties. Therefore, various approximations and simplifying assumptions are used to obtain solutions to the kinetic equation for specific physical conditions.

We note the basic physical mechanisms that determine the collection of energy by electrons and the formation of EES in the RFCD and are taken into account by various members of the kinetic equation. These are: 1) heating of plasma electrons by an RF field; 2) the electron heating processes associated with the NESCL, caused by: a) the electrons emitted by the electrode and accelerated by the fields; b) oscillation of the 'NESCL–plasma' interface. The latter mechanism has two varieties: 1) collisionless heating of plasma electrons by an oscillating layer boundary; 2) heating by an oscillating boundary of the layer in the presence of various electronic collisional processes.

The physical essence of the difference between the two last versions of the mechanism is that in the first case the electrons collide with the boundary of the layer and are reflected from it in the absence of elastic or inelastic collisions, and in the second – in the process of collisions with the layer, electrons experience many random collisions with the plasma particles.

There are other mechanisms that significantly complicate the process of formation of the EES of the RFCD plasma, for example, plasma instabilities, electronic collisions of the second kind, chemical reactions in molecular plasma, etc.

One of the most common simplifying assumptions is the neglect of the spatial variations of the applied electric field, which makes it possible to obtain the time evolution of the distribution function $f_e(\varepsilon)$ in the RFCD plasma (Fig. 5.1) [115].

In this case, two approaches are used:

1) the representation of the function $f_e(\varepsilon)$ in the form of two terms in the expansion in terms of harmonics in the velocity space and the Fourier expansion in time with substitution into the kinetic equation in order to obtain a system of equations that can then be solved numerically [257, 258]; 2) the solution of the kinetic equation by the Monte Carlo method. In the latter method trajectories of a large number of electrons are traced, as they are accelerated by a local electric field and dissipate and lose energy in elastic and inelastic collisions. In most Monte Carlo calculations of the characteristics of

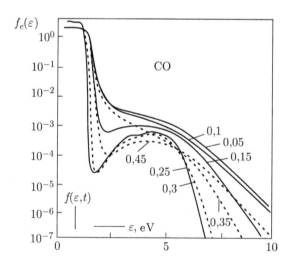

Fig. 5.1. The periodic behaviour of the electron energy distribution function $f_e(\varepsilon)$ as a function of $\omega/p = \pi 10^7$ s^{-1} · Torr^{-1} [115].

RF discharges, a spatio-temporal variation of the local electric field taken from the experiment was determined [259, 260].

Theoretical analysis shows that the behaviour of the function f_e (ε) within the period of the RF field is determined by the field frequency and physical conditions in the plasma [98]. Moreover, in not too strong spatially uniform fields, the frequency of the momentum relaxation v_{en} significantly exceeds the electron energy relaxation rate v_ε, which allows us to derive the kinetic equation for the symmetric part of the distribution function $f_0(\varepsilon, t)$ under the assumption $\omega \gg v_{en}$. Consideration of this kinetic equation shows: a) for $\omega \ll v_\varepsilon$, the function f_0 (ε, t) observes the quasi-stationary change in the field and is found from the solution of the stationary kinetic equation; b) in the case of $v_\varepsilon \ll \omega \ll v_{en}$, the function f_0 (ε, t) does not feel the field oscillations and corresponds to the root-mean-square value of the field E; c) for intermediate frequencies $\omega \sim v_\varepsilon$, the function $f_0(\varepsilon, t)$ is determined from the non-stationary kinetic equation and will depend not only on E, but also on ω; d) in the region of higher frequencies $\omega \sim v_{en}$ a separate special consideration is required.

In the implementation of the above approach to the solution of the kinetic equation, two variants were used [98]: 1) analytical calculation for the model collision integral, 2) numerical calculations for a particular gas.

If the dimensions of the non-uniformity region of the electric field in a plasma are comparable with the electron energy relaxation length l_ε or the relaxation time f_ε (ε) of the order of the RF field period, it is necessary to take into account the effect of spatial diffusion and electron drift on the form f_e (ε), and also the non-stationarity f_e (ε, t).

According to [96], the space-time non-locality of f_e (ε) is manifested when the following conditions are fulfilled:

$$\frac{1}{E}\left|\frac{dE}{dx}\right|\lambda\left(\frac{M}{m}\right)^{1/2} \geq 1, \quad \frac{m}{M}v_{en} \ll \omega \ll v_{en},$$

where λ is the electron mean free path. The numerical solution of the non-stationary kinetic equation in the spatially non-uniform field of the RF discharge showed a significant discrepancy with the calculations of f_e (ε) in the local approximation (Fig. 5.2). It is also shown here that the kinetic coefficients of the plasma (the mean electron energy $\langle\varepsilon_e\rangle$, the ionization rate constant $\langle\sigma_i v\rangle$), calculated without allowance and taking into account the non-locality of $f_e(\varepsilon)$, differ significantly.

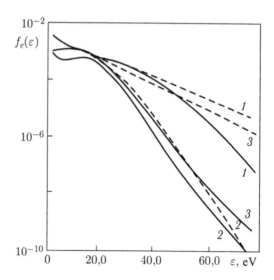

Fig. 5.2. The function $f_e(\varepsilon)$ in the RF discharge in He at the points of the space $x = 0$ (*1*), $x = 1.0$ cm (*2*), $x = 2.0$ cm (*3*). Solid curves for solving the non-stationary kinetic equation in the non-uniform field of the RF discharge; dotted lines are functions f_e (ε) calculated on the assumption of the spatial locality of the energy spectrum [96].

Due to the presence of excitation processes of vibrational energy levels, the process of formation of $f_e(\varepsilon)$ in a molecular plasma differs from that in atomic gases. In particular, it has been experimentally established that $f_e(\varepsilon)$ becomes localised in molecular gases at lower pressures ($p \geq 0.1$ Torr) than in atomic gases ($p > 5-15$ Torr) [261].

In the presence of appreciable populations of excited levels of atoms in the RFCD plasma, the form of the function f_e (*e*) and its time course are significantly influenced by collisions of the second kind of electrons with neutral particles, as a result of which a part of the energy lost by electrons in inelastic collisions of the first kind is returned back to them [115, 262]. In particular, the role of collisions of electrons with molecules in vibrational, electronic excited states in the RFCD plasma in the He–CO mixture was studied depending on the populations of vibrational CO ($v = 1$) and electronic CO (a^3P), He (2^3S) states (Fig. 5.3) [262].

From this it is clear that in the presence of superelastic collisions the number of medium-energy electrons increases noticeably, the time modulation of their density decreases, and the electron density in the 'tail' of the energy distribution increases by orders of magnitude throughout the entire RF field period. Due to collisions of the second kind, in the plasma of RFCD there is a tendency to acquire the properties of a DC discharge plasma.

A number of physical processes that affect the formation of the function $f_e(\varepsilon)$ and are manifested in the near-electrode plasma of the RFCD are investigated.

The most important of the above processes is the formation of near-electrode high-energy electron beams [47–50].

The energy distribution function of 'fast' electrons in the transverse RFCD was calculated in [118] with the help of the kinetic equation obtained on the basis of the following model representations: a monoenergetic beam formed from the electrode-generated electrons

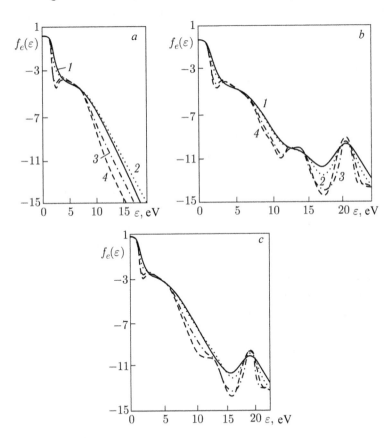

Fig. 5.3. Distribution functions $f_e(\varepsilon)$ in the plasma of the RF discharge in the 10–90% CO–H$_2$ mixture at different values of the parameter $t = \omega t/2np$ (curve 1 – $t = 0$, curve 2 – $t = 0.16$, curve 3 – $t = 0.25$, curve 4 – $t = 0.34$. The concentration of excited states: *a*) 0; *b*) CO ($v = 1$) / CO = 0.2; CO (a^3P) / CO = 10^{-6}; He (^3S)/He = 10^{-6}; *c*) CO ($v = 1$)/CO = 0.2; CO (a^3P)/CO = 10^{-4}; He (^3S)/He = 10^{-4} [262].

flies from the NESCL into the plasma; fast electrons are isotropically scattered on the atoms of the gas; an electron beam relaxing in space is considered equivalent to a source of electrons with an isotropic velocity distribution function and density

$$\rho(x,\varepsilon) = n_0 v_0 e^{-x/\lambda_0} \delta(\varepsilon - \varepsilon_0),$$

where v_0, n_0, λ_0, ε_0 are the collision frequency with atoms, the density of gas atoms, the mean free path and the energy of the primary electrons, respectively, $\delta(\varepsilon-\varepsilon_0)$ is the Dirac delta function; due to the small energy relaxation time of the 'fast' electrons, at each instant of time, $f_e(\varepsilon)$ is assumed to be stationary, depending only on the instantaneous voltage drop in the NESCL; in the near-electrode plasma there are no proper 'fast' electrons. The obtained form of the distribution function of electrons with energies greater than the first gas excitation potential ε_1 at various distances from the NESCL boundary is shown in Fig. 5.4.

In the presence of collisions of electrons with atoms, a non-monoenergetic near-electrode electron beam should form in the NESCL. Taking into account the different types of collisions in the NESCL when calculating the EES for the region of the space at the 'NESCL–plasma' boundary in the DCGD made it possible to obtain a detailed EES structure (Fig. 5.5) [263], which coincides with the experimental results under analogous conditions [264].

The calculation of the total EES in different phases of the RF field period in RFCD based on the solution of the non-stationary kinetic equation with allowance for the electron beam is given in [265]. It was assumed that there is a constant in time source of isotropically distributed beam monoenergetic electrons. The results are shown in

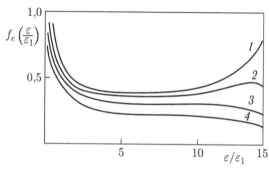

Fig. 5.4. The form of the distribution function $f_e(\varepsilon/\varepsilon_1)$ at various distances from the plasma boundary: $1 - x = 0$; $2 - x = r_{De}$; $3 - x = 2r_{De}$; $4 - x = 3r_{De}$. $\varepsilon_{eb} = 300$ eV [118].

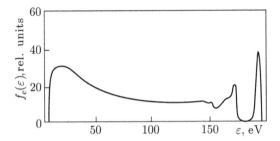

Fig. 5.5. The electron energy distribution function in the DCGD in the 'NESCL–plasma' boundary region [263].

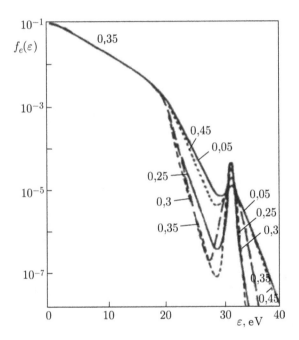

Fig. 5.6. The distribution function $f_e(\varepsilon)$. RFCD in He. $\omega/p = \pi \cdot 10^7$ s^{-1} · Torr^{-1}, $E/p = 6.0$ V · cm^{-1} · Torr^{-1}, the parameter t for the curves [265].

Fig. 5.6, illustrating the periodic in time behaviour of EES in the absence of a beam and in its presence. Hence it is clear that without a beam, the 'tail' of the EES strongly oscillates, and with the beam a noticeable tendency toward stationarity of the EES is expressed, especially in the vicinity of the value of the initial energy of the beam.

In [266], the PIC (particle-in-cell) method [267] was used to calculate the function $f_e(\varepsilon)$, formed by the oscillating NESCL of

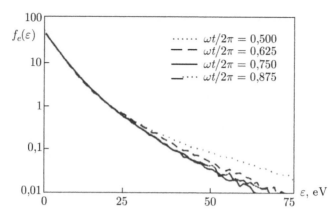

Fig. 5.7. The distribution function $f_e(\varepsilon)$ (in the region between 18 and 22 mm from the left electrode) at four different moments of the RF field period [266].

the RFCD for different phases of the RF field period (Fig. 5.7). The physical conditions were chosen in such a way that electron emission from the electrodes was neglected, and the dominant mechanism was collisional stochastic heating of plasma electrons. The energy distribution of electrons obtained for these conditions reveals a substantially 'long tail' modulated in time, characterized by a noticeable anisotropy of the motion of its constituent electrons.

The nature of the effect of the stochastic mechanism of electron heating on the low-pressure RFCD plasma was theoretically studied in [125]. It was found that it is important to take into account the distribution profile of the electric potential $\varphi(x)$ in a plasma that captures low-energy electrons that do not reach the periphery of a discharge with stochastic heating. Hence, the slow electrons that make up the bulk of the electron density in the plasma have a significantly lower diffusion energy coefficient than the high-energy electrons. Thus, there are two different energy scales of the electron distribution, and $f_e(\varepsilon)$ in such an RFCD is enriched both by slow and fast electrons [149]. It was also shown in [125] that in the following cases: 1) the spatially homogeneous ion distribution $n_i(x)$; 2) $\omega < \Omega_b = v_e/L_0$, where Ω_b is the transit frequency, v_e is the electron velocity, L_0 is the interelectrode distance; 3) $\omega > \omega_{0e}$ – stochastic heating disappears.

The influence of such factors as various plasma instabilities or chemical reactions on the mechanism of formation of plasma EES in RFCD is currently poorly understood.

The most applicable to the experimental study of the kinetics of low-pressure RFCD plasma electrons are the three groups of methods:

1) Langmuir's probe method [215]; 2) the electrostatic analyzer method [87, 246]; 3) optical–spectroscopic methods [102, 269, 270].

A significant variety of physical conditions encountered in RFCD can be covered only by the whole set of diagnostic methods that have limitations on the possibilities of their application.

The existence of anisotropic and time-varying constituents of the EESs in the RFCD, which requires an additional complication of the measurement procedure, seriously complicates the experimental obtaining of reliable information.

At present, various modifications of the probe method are most often used to determine the function $f_e(\varepsilon)$. At the same time, it must be emphasized that this method can be used to investigate a plasma with only a very narrow EES of the order of several tens of eV. When the voltage is applied to the probe $V_p \geq 100$ V the plasma under investigation is unacceptably indignant. As a result, only the EES of the positive column plasma is determined by the probe method, or only with a rather limited accuracy – the groups of slow electrons of the function $f_e(\varepsilon)$ with a wide energy interval (for example, in the near-electrode plasma).

In a number of works, the EES of a near-electrode plasma was determined with the help of an electrostatic analyzer of charged-particle energies, located, as a rule, behind the grounded RFCD electrode. A significant drawback of this method is the presence between the plasma and the analyzer of the charged-particle energies of the near-electrode SCL with a complicated, time-varying complex of physical processes.

Great opportunities in studying the electronic kinetics of the plasma of the low-pressure RFCD are offered by the spectral–optical methods, for example, the rapidly progressing method of polarization spectroscopy, which, in particular, provides unique information on the anisotropy of the kinetic processes in plasma [269].

Unfortunately, until now the contactless method for reconstructing the distribution function $f_e(\varepsilon)$ through the optical spectrum of plasma radiation by solving the inverse ill-posed problem by the regularization method has not been properly developed [270]. Satisfactory development of this promising technique is restrained by a number of reasons, in particular, the need to have an extensive bank of reliable spectroscopic information about the gases under investigation, high requirements for spectroscopic measurement systems with an abnormally large dynamic range, and so on.

Even simpler optical methods, in combination with contact methods, provide a significant amount of information on the EES of the RFCD plasma [102].

Despite the unconditional dependence of the EES of the RFCD plasma on time, the overwhelming majority of studies are devoted to determining the quasi-stationary form of this plasma characteristic. The latter is due not only to procedural reasons, but also to the value of information about the physical conditions in the plasma of the discharge under investigation, which is contained even in the time-averaged distribution function $f_e(\varepsilon)$ of the electrons. The latter applies, in particular, to the basic group of 'slow' electrons, whose behaviour often depends little on time.

Experimental studies show that when the parameters of RFCD are changed in a wide range, the EES of its plasma undergoes significant changes. Moreover, in order to study in more detail the processes of electron kinetics, it is expedient to distinguish two types of plasma:

1) with a relatively narrow EES ($0 < \varepsilon_e < 10^2$ eV);

2) with a wide EES ($0 < \varepsilon_e < 10^4$ eV).

Until now, the object of research was mainly the first type of plasma of RFCD, and the second one remains poorly studied.

Plasma with a narrow EES appears in α-discharges, as well as in γ-discharges in the region – an analog of the plasma of the positive column in DCGD. A wide EES is observed in the near-electrode plasma of γ-discharges, which is analogous to the region of 'negative luminescence' of DCGD.

In a number of studies, the character of the function $f_e(\varepsilon)$ of the first type and the mechanisms of its formation have been investigated by the probe method [113, 119, 259].

The form of $\ln f_e(\varepsilon)$ at the centre of the RFCD in He is shown in Fig. 5.8 [119].

From this it is clear that as the gas pressure increases from 0.1 to 2 Torr, the function $f_e(\varepsilon)$, varying from a single Maxwellian distribution, gradually turns into a distribution with two quasi-Maxwell groups of electrons. Under these conditions, the high-energy 'tail' of EES can be enriched both by accelerated γ-electrons and by the oscillating boundary of the NESCL. At a pressure of 0.1 Torr, the mean free path of electrons λ_e is close to the interelectrode distance d. Therefore, the plasma electrons are more often elastically colliding with the oscillating boundary of the NESCL than with neutral atoms, experiencing on average 'stochastic acceleration' [56]. As a result, the plasma electrons heated by the above mechanisms energetically

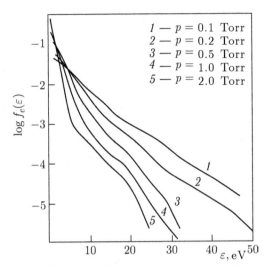

Fig. 5.8. Dependence of the distribution function $f_e(\varepsilon)$ on RF voltage. RFCD in helium. Power $P = 20$ W, $f = 27$ MHz. $V_{\sim}$: $1 - 250$ V, $2 - 240$ V, $3 - 225$ V, $4 - 200$ V, $5 - 175$ V [119].

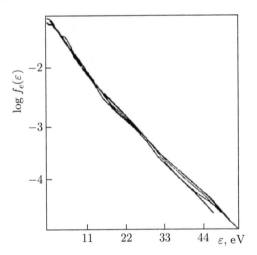

Fig. 5.9. Dependence of the distribution function $f_e(\varepsilon)$ on the distance to the active electrode (from 7.5 to 25 mm). RFCD in helium. $p = 0.1$ Torr, $f = 27$ MHz, $V_{\sim} = 250$ V [119].

relax to a quasi-Maxwellian $f_e(\varepsilon)$ plasma in equilibrium with the RF field in the plasma.

An investigation of the behavior of the function $f_e(\varepsilon)$ in space (Fig. 5.9) confirms the expected homogeneity of the conditions in spatial scales of the order of λ_e, despite the asymmetry in the discharge caused by the active (not grounded) electrode as the preferred source of electron heating.

A different picture is observed for $p = 2$ Torr, when $\lambda_e \sim 1$ mm and for the order of the width of the transition region 'NESCL–plasma'. This corresponds to the 'wave-riding' mode of heating of plasma electrons (along with heating by γ-electrons) in inelastic collisions of electrons with an oscillating boundary of the NESCL under conditions of frequent collisions with neutral atoms. The spatial dependence of $f_e(\varepsilon)$ for this case is shown in Fig. 5.10. It is clear from this that at all the investigated distances from the electrode the function $f_e(\varepsilon)$ is non-Maxwellian, enriched by 'fast' electrons, with two characteristic quasi-Maxwellian groups of electrons with different effective 'temperatures'. In this case, the high-energy 'tail' $f_e(\varepsilon)$ rapidly decreases as it moves away from the active electrode and the oscillating boundary of the NESCL.

In [119], a dependence on the type of gas in the behaviour of the function $f_e(\varepsilon)$ at different pressures was also found. Thus, the effect of collisions with neutral atoms on the kinetics of electrons in RFCD, in Ar begins to appear at pressures that are one order of magnitude lower than in He, which is due to the large difference in the cross sections for electron collisions with the atoms of these gases.

The character of the change in the function $f_e(\varepsilon)$ at the centre of the symmetric RFCD with a smooth transition from the α-mode to

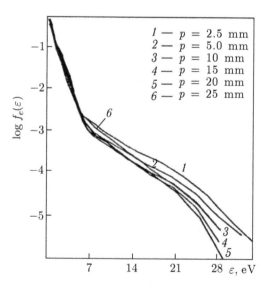

Fig. 5.10. Dependence of the distribution function $f_e(\varepsilon)$ on the distance to the active electrode. He, $p = 2$ Torr, $V_- = 175$ V [119].

the γ-mode of the discharge in Ar and He was investigated by the improved probe method in [116]. The measurements were made at a fixed gas pressure $p = 0.3$ Torr and a field frequency of 13.56 MHz in a relatively wide range of RF voltages $V_{\sim} = 20{-}700$ V.

The evolution of the function $f_e(\varepsilon)$ with the change in the RFCD parameters for Ar is shown in Fig. 5.11, and for He $-$ in Fig. 5.12.

As can be seen from this, for small discharge currents, the functions $f_e(\varepsilon)$ do not change significantly with increasing current,

remaining in Ar almost Druvesteyen $\left(f_e(\varepsilon) \sim \exp\left(\dfrac{\varepsilon_e}{ekT} \right)^2 \right)$ and in He

$-$ quasi-Maxwellian $\left(f_e(\varepsilon) \sim \exp\left(\dfrac{\varepsilon_e}{ekT} \right) \right)$, where T is the parameter equal to the electron temperature for the Maxwellian distribution.

The difference in the behavior of $f_e(e)$ for Ar and He is due to a different dependence on the electron energy of the frequency of the elastic collisions of electrons with atoms for these gases.

When a certain value of the discharge current (RF voltage) is reached, an $\alpha \rightarrow \gamma$-change in the RFCD modes occurs with a sharp increase in the electron concentration and a decrease in the effective electron temperature.

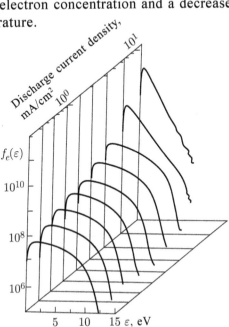

Fig. 5.11. Dependence of the distribution function $f_e(\varepsilon)$ on the discharge current density. RFCD in argon. $p = 0.3$ Torr [116].

A sharp drop in $T_{e\,\mathrm{eff}}$ and an increase in n_e lead to a jump-like increase in the frequency of electron–electron collisions $v_{ee} \sim n_e / T_e^{3/2}$, which effectively intensifies the process of Maxwellization of low-energy electrons in both gases.

In the RFCD of the γ-mode in He the measured $f_e(\varepsilon)$ detects the 'tail' of high-energy electrons and can be represented as the sum of two Maxwellian distributions with temperatures of the group of 'cold' electrons T_{ec} and the group of 'hot' electrons T_{eh} (Fig. 5.12). The measured temperature T_{ec} is much higher than that calculated on the basis of the local balance of the electron energy, taking into account ohmic heating and losses in elastic collisions. This difference can be explained by the additional process of heating the 'slow' electrons by transferring energy from the group of high-energy electrons through Coulomb collisions, which occurs, for example, in the region of 'negative luminescence' of DCGD [271].

The situation with the study of an even simpler case of the RFCD with a narrow EES is not satisfactory. Practically the only probe diagnostics used provides reliable information about a very narrow range of electron energies, does not allow analyzing the behaviour of the 'tails' $f_e(\varepsilon)$ with increased energies, and the determination

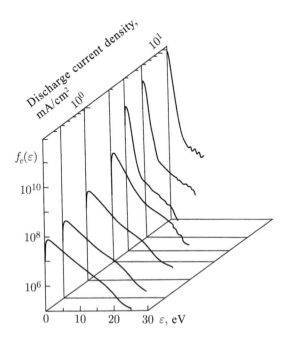

Fig. 5.12. Dependence of the distribution function $f_e(\varepsilon)$ on the discharge current density. RFCD in helium. $p = 0.3$ Torr, $f = 13.56$ MHz [116].

of the real boundary of the plasma EES. The directional component of the EES of the plasma of RFCD is practically not investigated.

Thus, the high-energy part of the function $f_e(\varepsilon)$, which is responsible for inelastic processes in the plasma, remains experimentally poorly studied. The mechanism of formation of two autonomous stable quasi-Maxwellian groups of electrons in the RFCD plasma, the nature of their interaction, requires additional study.

The plasma of RFCD with a wide EES has been studied to a much lesser degree.

Of the early studies of this kind, only two can be mentioned. The first is devoted to a systematic study with the help of probes and an energy analyzer of EES of moderate width ($0 < \varepsilon_e \leq 200$ eV) [47, 48, 50], the results of which are presented below. In the second work [53] of technological direction with the help of an energy analyzer, the fact of the presence in the plasma of RFCD of electrons with energies $0 < \varepsilon_e \leq 2$ keV was discovered but the mechanism of their appearance was not explained in any way (Fig. 5.13).

The measurements carried out with the energy analyzer also confirmed the possibility of the above situation: on the delay curves, the 'tail' of electrons, whose energy reached 1.1 keV, was observed, which is equivalent to the energy of twice the applied RF voltage [88]. The type of EES is not given in this paper, but it is noted that the current of fast electrons decreases practically linearly with increasing delay, indicating that $f_e(\varepsilon)$ is constant over a wide energy interval.

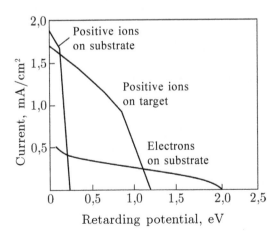

Fig. 5.13. RFCD. The current of positive ions and electrons as a function of the retarding potential in the energy analyzer on a grounded electrode [53].

The time course of the EES of the plasma of the low-pressure RFCD within the RF field period is still an experimentally little studied one. There are only some papers devoted to measurements with time resolution [89, 110].

The known trends in the temporal behaviour of EES according to the ratio of the frequency of the field ω with the characteristic frequencies in the plasma – the ambipolar frequency $v_a = \tau_a^{-1}$ (τ_a is the ambipolar lifetime of charged particles), the energy exchange frequency for elastic collisions δv_{en} ($\delta = 2m/M$) and the frequency of inelastic collisions of electrons with atoms v^* are as follows [272].

Under the condition $\omega < v_a$, both the electron concentration n_e and $f_e(\varepsilon)$ are determined by the instantaneous values of the external parameters of the discharge. In the case $\omega > v_a$, the concentration n_e does not have time to vary over the field period and only the form of the function $f_e(\varepsilon)$ depends on time. Usually in the conditions under consideration $v^* \gg \delta v_{en}$. Then we have: a) $v_a \ll \omega \ll \delta v_{en}$ – the entire function $f_e(\varepsilon)$ changes with time; b) $\delta v_{en} \ll \omega \ll v^*$ – the low-energy part of the EES to the energy threshold of inelastic collisions does not depend on time, only the 'tail' of the distribution function is modulated; c) $\omega \gg v^*$ – EES modulation is absent.

Time-resolved measurements of EES were carried out using a special electronic circuit with a Langmuir probe in the central region of the low-frequency (100–400 kHz) RFCD in Ar at a pressure of 0.2 Torr in [110]. The experimental results obtained are shown in Figs. 5.14 and 5.15.

The parameters of the low-energy group vary slightly with time. The high-energy group of electrons varies significantly during the RF field period, beginning with energies corresponding to the threshold values of the energies of inelastic electron collisional processes. The density of fast electrons reaches a maximum in the middle of the negative half-cycle of the RF field. As will be shown below, the probe method gives unsatisfactory information about the 'tail' of the distribution function. In fact, the energy range of the EES under the conditions under consideration is much wider. As can be seen from the obtained functions $f_e(\varepsilon)$, it is completely unclear where the 'tail' of the distribution ends. It is not obvious that the EES consists of only two groups, there must still be a third group of electrons in the vicinity of the energies of the γ-electron beams.

Measurements with a temporal resolution of the EES of the electrons bombarding the grounded electrode were carried out in [89] with the help of a gated power analyzer of the Yuza–Rozhansky

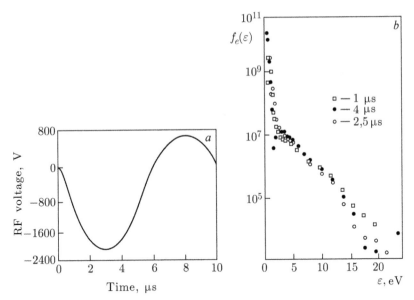

Fig. 5.14. RFCD in argon. The form of the RF voltage (*a*) (frequency 100 kHz) and (*b*) the time-resolved function $f_e(\varepsilon)$ measured at three different moments of the RF field period at the centre of the discharge gap [110].

type with a gate duration of 10 ns in a low-frequency RFCD in Ar at a field frequency of 13.56 MHz. The fixed energy distributions of the electron currents in the analyzer for different time delays τ (phases of the RF field period) are presented for two pressures of 0.004 and 0.018 Torr in Fig. 5.16.

Here, $\tau = 0$ corresponds to the moment of maximum approximation of the boundaries of the 'plasma–layer' boundary to the grounded electrode; $\tau < 0$ – motion of the layer boundary to the electrode; $\tau > 0$ ' the motion of the boundary from the electrode. For $\tau = 0$ (Figs. 5.16 *b*, *e*) two groups of electrons are observed: slow electrons with energies up to 20 eV and fast electrons with energies up to 1 keV.

The EES of fast electrons at $\tau = 0$ have a 'triangular' shape with a maximum energy close to twice the amplitude value of the applied RF voltage. For $\tau < 0$ (Fig. 5.16 *a*, *d*), the spectrum expands toward low energies. Moreover, at a pressure of 0.18 Torr, the shape of the spectrum remains 'triangular', and at a pressure of 0.004 Torr it becomes an almost horizontal 'plateau'.

The appearance of wide EES at $\tau < 0$ is attributed to the effect of spatial focusing of electrons of different energies [89]. However, this effect in this case should be very weakly expressed, since the

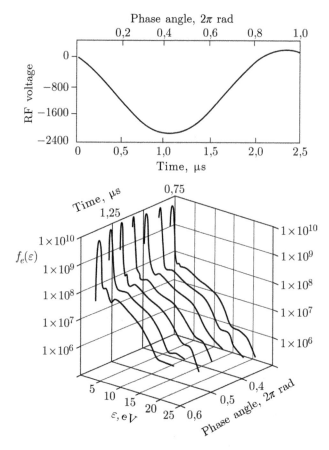

Fig. 5.15. RFCD in argon, $p = 0.2$ Torr, power $P = 150$ W. The form of the RF voltage (*a*) (frequency 400 kHz) and (*b*) the time-resolved function f_e (ε) measured at three different moments of the RF field period at the centre of the discharge gap [110].

transit times of the discharge gap by electrons with energies of 500-1000 eV are very small and close even for electrons with boundary values of the energy range. Apparently, here the Langmuir waves are excited by an electron beam, which is effectively scattered by the kinetic instability of the beam observed in the RFCD in Ar at a pressure of 0.02 Torr [245]

Summarizing the above, it should be noted that in the literature there are papers devoted to experimental studies of RFCD with large area electrodes, when the density of the near-electrode electron beam is small and when determining the EES, the primary electrons of the beams were not fixed. Moreover, these studies were carried out, as a rule, in the central regions of RFCD, where electron beams have already been relaxed in momentum and energy.

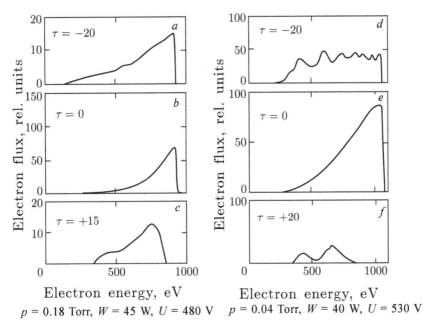

Electron flux, rel. units — Electron energy, eV

$p = 0.18$ Torr, $W = 45$ W, $U = 480$ V $p = 0.04$ Torr, $W = 40$ W, $U = 530$ V

Fig. 5.16. The distribution function $f_e(\varepsilon)$, measured at different times of the RF field [89].

Practically in all known papers related to RFCD, there are theoretical or experimental data for narrow EES with electron energies $\varepsilon_e \le 50$ eV.

5.4.2. Statement of the problem of our research

The formulation of the problem of our investigations described below is due to the following considerations.

An experimental study of the mechanism of the formation of the EES of near-electrode low-pressure RFCDs, its dependence on the external discharge parameters, the geometry of the discharge gap for the development of methods for creating plasma sources with an electron energy distribution controlled in a wide range was the main work.

The focus was on the study of the near-electrode plasma in the vicinity of a grounded electrode of an asymmetric RFCD with increased values of the RF voltage supporting the discharge.

The EES of the near-electrode plasma should occupy a large energy interval, since in the NESCL of the RFCD under investigation almost all the drops of the RF and quasi-stationary fields present in the discharge gap are concentrated.

The investigated near-electrode plasma near an grounded electrode is created by an electron beam from the active electrode and represents a practically unexplored variety of gas-discharge plasma. The presence of a grounded electrode makes it possible to significantly vary the physical conditions in the near-electrode plasma, using V_0- and I_0-modes of RFCD. In the I_0-mode, as is known, the parameters of the NESCL at the grounded electrode are weakly modulated in time, and in the V_0-mode, the NESCL experiences a periodic collapse with the frequency of the rf field.

In the studies described below, the V_0-mode of RFCD was predominantly investigated, which is due to the following reasons: 1) the presence of the dynamics of physical processes in the NESCL; 2) the possibility of more efficient use for measurements of the energy analyzer of charged particles; 3) investigation of near-electrode plasma under conditions close to those in technological processes.

The object of the study was the RFCD in He, Xe and air at pressures $p = 10^{-2}-1$ Torr in the frequency range of the field 0.5–5 MHz, the amplitudes of the RF voltage $V_{\sim} = 100-2000$ V.

The choice of working gases is determined by the desire to elucidate the general laws governing the formation of EES in the plasma of atomic and molecular gases, among which He is most often used, especially in various mixtures of gases as a buffer gas, and a universal gas such as air.

The study examined the low-density RFCD as a result of the relevance of its use in the plasma chemical technology of modifying the physical properties of surfaces, gas lasers, plasma chemistry, and in connection with a detailed study of the physical mechanism of RFCD. Precisely under these conditions, the near-electrode voltages of quasi-stationary electric fields, the energy of electron beams increase, intensive Langmuir plasma oscillations are excited, i.e., factors forming highly non-equilibrium EES of the plasma under investigation are excited.

The use of predominantly high RF voltages ($V_{\sim} \geq 1$ kV) is explained by the greatest interest in the investigation of RFCD in the γ-mode and the desire to study the little-studied processes of formation of wide EES in the near-electrode plasma of RFCD.

The selected section of the radiofrequency band of the electric field frequencies is due to the fact that at such frequencies: 1) there is a phenomenon of RF detection, providing significant quasi-stationary near-electrode voltages; 2) it is easier to carry out of a time-resolved study of physical processes in RFCD; 3) the level of electrical interference in the measuring systems and parasitic capacitances in the electrical circuits of the discharge decreases; 4) according to preliminary data, all the main elements of the physical mechanism of RFCD, inherent to the latter in the case of higher frequencies of the RF field (f = 13.56 MHz), which are most often used in practice, are retained.

Experimental study of the EES of the RFCD plasma is not a simple task and is complicated by the fact that the existing methods of diagnosis are not free of shortcomings. Therefore, it is desirable to simultaneously use alternative methods to increase the degree of reliability of the results obtained. The following diagnostic methods were used in the work:

1) graphic processing of Langmuir probe characteristics [273]; 2) the method of automatic measurement of the second derivative of the probe current $d^2I\,(V)/dV^2$ [215]; 3) Measurement of the energies of plasma electrons by means of an energy analyzer [87, 246, 274].

5.4.3. Electronic energy spectrum of RFCD plasma when the condition $\omega_{e0} \leq v_{en}$ is satisfied

Let us consider the results of an experimental study of the electron energy distribution in the RFCD plasma under conditions of a significant role of collisional processes.

At the same time, RFCD in He was studied at a pressure p = 0.5 Torr in a discharge tube with electrodes 5 cm in diameter and an interelectrode distance d = 10 cm.

The dependence of the distribution function $f_e\,(\varepsilon)$ on the amplitude of the RF voltage in the central region of the discharge is shown in Fig. 5.17.

The change in the character of the energy distribution of electrons along a sufficiently extended discharge gap also studied is shown in Fig. 5.18.

Let us analyze the physical conditions in such a discharge.

The plasma parameters have the following values: charge density n_e = 10^9–10^{10} cm^{-3}; the effective temperature of the 'slow' electrons $T_{e1} \leq 10$ eV; the concentration of atoms is $n_a \sim 10^{16}$ cm^{-3}.

The characteristic electron energy of the near-electrode electron beams is $\varepsilon_{eb} \sim (1-5) \cdot 10^2$ eV.

The penetration depth of the beam into the plasma [128]

$$\lambda = \left(\frac{1}{3} \frac{\varepsilon_{eb}}{\varepsilon_0} \lambda_{ib}\, \lambda_{bn} \right)^{1/2} \,, \text{ where } \lambda_{ib}, \lambda_{bn} \text{ is the ionization length and}$$

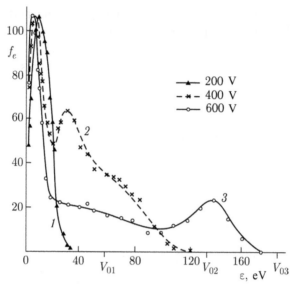

Fig. 5.17. Dependence of the distribution function $f_e (\varepsilon)$ in the central region of RFCD on RF voltage. RFCD. He, $p = 0.5$ Torr, $f = 8$ MHz, the distance from the electrode to the probe is $d_{e-p} = 40$ mm. $V_\sim = 200, 400, 600$ V.

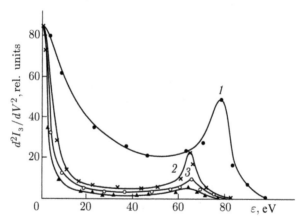

Fig. 5.18. The change in the character of the electron distribution along the energy gap along the discharge gap. RFCD. He, $p = 0.5$ Torr, $f = 10$ MHz, $V_\sim = 280$ V. The distances between the electrode and the probe are d_{e-p}: *1* – 14 mm, *2* – 76 mm, *3* – 85 mm, *4* – 88 mm.

length free path of the beam electrons in the gas, respectively, and ε_0 is the effective energy loss in the ionizing collision.

For the conditions under consideration, we obtain $\lambda_{eb} = 0.8$–1.8 cm $\ll d$.

The relaxation length of plasma electrons with respect to energy is $l_\varepsilon \sim 3 \cdot 10^2$ cm $\gg d$.

The distributions in Fig. 5.18 also show that the non-equilibrium $f_e(\varepsilon)$ caused by the beam relaxes relatively slowly in space, although somewhat faster than in estimating l_e.

Let us estimate the ratio of frequencies ω_{0e} and v_{en} for these conditions: $\omega_{0e} = 1.8 \cdot 10^9$ s^{-1}, $v_{en} = 1.2 \cdot 10^9$ s^{-1}, i.e. $\omega_{0e} \approx v_{en}$.

The characteristic time of development of the beam instability [155]:

$$\tau_0 \sim \frac{1}{\omega_{e0}}\left(\frac{2n_e}{n_{eb}}\right)^{1/3} = \frac{1}{1.8 \cdot 10^9}\left(2 \cdot \frac{10^9}{10^6}\right)^{1/3} = 7 \cdot 10^{-9} \text{ s},$$

where $n_e/n_{eb} \sim 10^3$ was assumed.

The time between two collisions of an electron with atoms

$$\tau \sim \frac{1}{v_{en}} = \frac{1}{1.2 \cdot 10^9} = 0.8 \cdot 10^{-9} \text{ s}.$$

Thus, we obtain: $\tau_0 \gg \tau$, which means that it is impossible to develop a beam instability.

It follows that the beam is scattered only in collisions with neutral atoms, enriching the plasma distribution function $f_e(\varepsilon)$ by 'fast' electrons.

Let us consider the peculiarities of the balance of charged particles in an RF discharge plasma with such a non-Maxwellian function $f_e(\varepsilon)$. We use for this purpose the experimentally obtained function $f_e(\varepsilon)$, shown in Fig. 5.17. As can be seen from this figure, the energy distribution in the group of 'slow' electrons is well approximated by the Maxwellian distribution with $T_e = 5.4 \cdot 10^4$ K.

Under the conditions considered here, we estimate the ionization frequencies due to the obtained distribution function $f_e(\varepsilon)$ and directly due to the RF field present in the plasma. Moreover, we estimate the ratio of the contributions to the ionization of the group of 'slow' electrons with the Maxwellian energy distribution ($T_e = 5.4 \cdot 10^4$ K) and the remaining electrons of the considered distribution function, the 'fast' electrons. Let us denote the average ionization frequencies due to 'slow' and 'fast' electrons through $\overline{Z_i(T_e)}$ and $\overline{Z_i^{\text{fast}}}$, respectively.

Then, for the 'slow' and 'fast' electrons, the construction of the function $f_e(\varepsilon) \cdot P_i(\varepsilon) \sim \overline{Z_i}$ (where $P_i(\varepsilon)$ is the ionization cross section for electrons with energy ε) will allow us to find the ratio of the

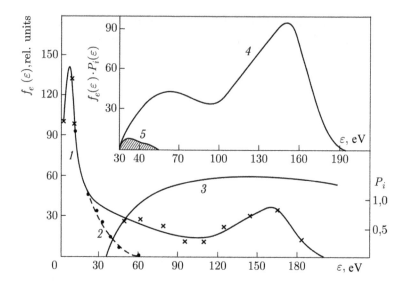

Fig. 5.19. Determination of the relative contribution of the groups of 'slow' and 'fast' electrons to the ionization process for a non-Maxwellian distribution function $f_e(\varepsilon)$ in the RFCD plasma. *1* is the distribution function $f_e(\varepsilon)$, *2* is the approximation of the group of 'slow' electrons by a Maxwellian distribution with a temperature $T_e = 54 \cdot 10^3$ K, *3* is the curve of the dependence of the ionization probability $P_i(\varepsilon)$ of He atoms on the electron energy, *4* is the dependence of the quantity $[f_e(\varepsilon) \cdot P_i(\varepsilon)]$, proportional to the ionization frequency Z_i, on the electron energy, *5* is the contribution to the value $[k(\varepsilon) \cdot P_i(\varepsilon)]$, electrons with the Maxwellian distribution ($T_e = 54 \cdot 10^3$ K, curve *2*).

contributions of these groups to ionization as the ratio of the areas bounded by these curves (Fig. 5.19).

It follows from Fig. 5.19, 98% of the contribution to the ionization frequency $\overline{Z}_i$, conditioned by the function $f_e(\varepsilon)$ under consideration, is ionization by 'fast' electrons, i.e., $\overline{Z_i^{\text{fast}}} \approx 50\overline{Z}_i(T_e)$.

The absolute value of the quantity $\overline{Z}_i(T_e)$ is determined by the formula [182]:

$$\overline{Z}_i(T_e) = 600\left(\frac{2e}{\pi m}\right)^{1/2} \cdot apV_i^{3/2} \frac{\left(\dfrac{kT_e}{eV_i}\right)}{\exp\left(\dfrac{eV_i}{kT_e}\right)}.$$

where $a = 4.6 \cdot 10^{-2}$ ion pairs/V $\cdot$ cm $\cdot$ Torr, V_i is the ionization potential of the working gas.

We have

$$\overline{Z_i}(T_e) = 600 \left(\frac{2 \cdot 4.8 \cdot 10^{-10}}{3.14 \cdot 9 \cdot 10^{-28}} \right)^{1/2} \cdot 4.6 \cdot 10^{-2} \cdot 0.5 \cdot \left(\frac{24.5}{3 \cdot 10^2} \right)^{3/2} \times$$

$$\times \frac{\left(\dfrac{1.4 \cdot 10^{-16} \cdot 5.4 \cdot 10^4 \cdot 300}{4.8 \cdot 10^{-10} \cdot 24.5} \right)^{1/2}}{\exp\left(\dfrac{4.8 \cdot 10^{-10} \cdot 24.5}{300 \cdot 1.4 \cdot 10^{-16} \cdot 5.4 \cdot 10^4} \right)} \approx 4 \cdot 10^5 \, \text{s}^{-1}.$$

From this $\overline{Z_i^{\text{fast}}} \approx 2 \cdot 10^7 \, \text{s}^{-1}$.

Let us now estimate the ionization frequency $\overline{Z_i}$ (RF) due to the RF field. In this case, according to the experimental data, we take $E_\sim \approx 10 \, \text{V} \, / \, \text{cm}$.

According to [157]:

$$\overline{Z_i}(\text{RF}) = \alpha v_d,$$

where α is the Townsend volume ionization coefficient, v_d is the electron drift velocity.

We have, further: $v_d = b_e E_\sim$, $b_e = e/m v_{en}$ – mobility of the electron. For our conditions:

$$b_e = \frac{4.8 \cdot 10^{-10}}{9 \cdot 10^{-28} \cdot 10^9} = 5.3 \cdot 10^8 \, \text{cm}^2 \, / \, (\text{V·s}),$$

$$v_d = 5.3 \cdot 10^8 \cdot \frac{10}{300} = 17.6 \cdot 10^6 \, \text{cm} \, / \, \text{s}, \quad \frac{E_\sim}{p} = \frac{10}{0.5} = 20 \, V \, / \, (\text{cm·Torr}).$$

According to the data of [157]: for the value $E/p = 20$ V/(cm · Torr), the quantity $\alpha/p = 2 \cdot 10^{-1}$ ion pairs / (cm · Torr).

Hence: $\alpha = 2 \cdot 10^{-1} \cdot 0.5 = 0.1$ ions/cm. Thus, we obtain:

$$\overline{Z_i}(\text{RF}) = 0.1 \cdot 17.6 \cdot 10^6 = 17.6 \cdot 10^5 \, \text{s}^{-1}.$$

As a result we have:

$$\overline{Z_i}(T_e) = 4 \cdot 10^5 \, \text{s}^{-1}; \quad \overline{Z_i}(\text{RF}) = 17.6 \cdot 10^5 \, \text{s}^{-1};$$
$$\overline{Z_i^{\text{fast}}} = 200 \cdot 10^5 \, \text{s}^{-1}.$$

Consequently, the main contribution to ionization in this case is caused by ionization due to 'fast' electrons.

As was established earlier, the balance of charged particles in such a plasma is maintained by their diffusion treatment with an anomalously large 'effective' diffusion coefficient, D_a, which is also due to 'fast' electrons.

The curves in Fig. 5.17 of the distributions $f_e(\varepsilon)$ show that, with increasing RF voltage the EES of the plasma expands. The temperature T_{e1} of the group of 'slow' electrons decreases monotonically. The latter, according to [119], indicates the decisive role of collisional processes in the formation of EES. In particular, this experimental fact shows that the effect of stochastic heating of electrons was absent.

The condition $\omega_{0e} \ll v_{en}$ was performed in experiments with a transverse RFCD of medium pressure.

5.4.4. EES plasma of RFCD in the case of $\omega_{0e} \gg v_{en}$

The conducted studies of the EES (electronic energy spectrum) of the near-electrode plasma in RFCD in He, Xe, and air at lower gas pressures ($p \leq 10^{-1}$ Torr) in discharge tubes. In these experiments, the condition $\omega_{0e} \gg v_{en}$ was satisfied, which implied a decrease in the role of the collision processes in the plasma and the possibility of the appearance of new effects of the beam–plasma interaction.

The EES of the RFCD plasma in He, depending on the applied RF voltages, is shown in Fig. 5.20.

The results of the study of EES in the RFCD in Xe for two field frequencies are shown in Figs. 5.21, 5.22.

The EES of the RFCD plasma in air for different RF voltages is shown in Fig. 5.23.

First, let us consider some characteristic features of the obtained EES of the RFCD plasma with plane symmetric electrodes.

Due to the nature of the work of the energy analyzer, not all of the EESs were recorded: the slowest plasma electrons did not enter the energy analyzer and, on the other hand, high-energy electrons, due to their low density, were not recorded by the measuring system. Experimental EESs included electrons with energies $\varepsilon_e \leq 60$ eV.

The obtained quasi-stationary EESs of the RFCD plasma in He are close to those measured by the probe method for the energy distributions of electrons in the central regions of similar RFCDs, given in Refs. [119, 259].

All these studies are characterized by the presence of a 'tail' of high-energy electrons, which is not fixed, however, satisfactorily due to experimental difficulties, and also by the presence of 1–2 groups of electrons in the EES of the plasma, depending on the discharge parameters.

With the increase in RF voltage, the previously observed trends can be traced in the obtained EES: 1) the energy interval of EES is expanded [50]; 2) the value of the mean energy $\bar{\varepsilon}_e$ of the group of 'slow' electrons decreases [50]; 3) there is a transformation of EES with two groups of electrons into a quasi-Maxwellian single group of electrons [259].

Let us analyze the mechanisms responsible for the formation of the obtained quasi- stationary EES of the RFCD plasma.

The study of the EES of near-electrode plasma near a grounded electrode is of considerable interest, since the mechanism of spectrum formation can include several physical processes simultaneously. This gives great opportunities for a focused formation of the plasma EES in a wide range of energies.

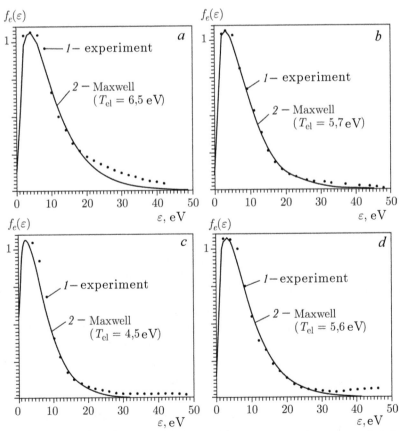

Fig. 5.20. Electron distribution functions by energy. RFCD. He, p = 0.1 Torr, f = 1.2 MHz. *a)* $V_\sim$ = 600 V, *b)* $V_\sim$ = 800 V, *c)* $V_\sim$ = 1000 V, *d)* $V_\sim$ = 1200 V.

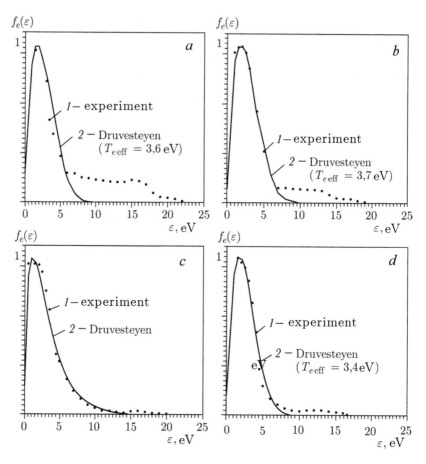

Fig. 5.21. Electron distribution functions by energy. RFCD. Xe, p = 0.1 Torr, f = 0.6 MHz. *a*) $V_\sim$ = 800 V, *b*) $V_\sim$ = 1000 V, *c*) $V_\sim$ = 1200 V, *d*) $V_\sim$ = 1400 V.

Because of the large relaxation length of the energy distribution function of the electrons ($l_\varepsilon > d$), physical processes occurring in the vicinity of both grounded and active electrodes should take part in the formation of the EES of the plasma near the grounded electrode.

Let us enumerate the corresponding possible physical processes in RFCD: 1) heating of plasma electrons by an RF field of high intensity in the vicinity of the active electrode; 2) enrichment of the plasma near the grounded electrode by high-energy electrons of the electron beam from the active electrode as a result of collisions with neutral particles; 3) plasma enrichment by higher energy electrons due to scattering of a relatively low-energy electron beam from a grounded electrode on neutral particles; 4) scattering of electron beams from both electrodes by Langmuir oscillations excited in

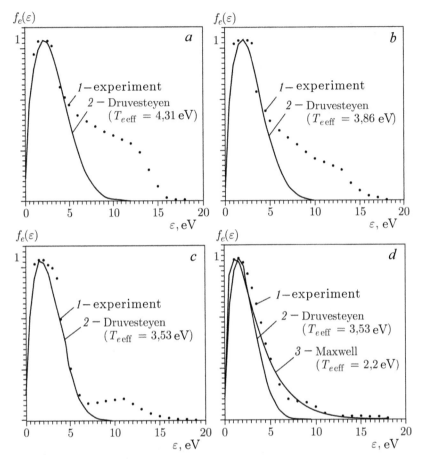

Fig. 5.22. Electron distribution functions by energy. RFCD. Xe, $p = 0.1$ Torr, $f = 1$ MHz. a) $V_\sim = 600$ V, b) $V_\sim = 800$ V, c) $V_\sim = 1000$, d) $V_\sim = 1200$ V.

the near-electrode plasma; 5) heating of plasma electrons by fields of beam instability; 6) exchange of electron energies of different energy groups as a result of their Coulomb collisions at high plasma electron densities; 7) stochastic heating of the plasma electrons by the boundary of the near-electrode potential barrier oscillating with the frequency of the RF field; 8) enrichment of plasma by electrons of elevated energies as a result of the appearance in the RFCD of the wall electron beams in the vicinity of the electrodes.

The contribution of each of the above processes to the mechanism for maintaining RFCD depends on the discharge parameters.

In these studies, RFCD was studied in three different gases: 1) a light inert gas He; 2) a heavy inert Ramsauer gas Xe; 3) mixtures of molecular gases N_2 and O_2–air.

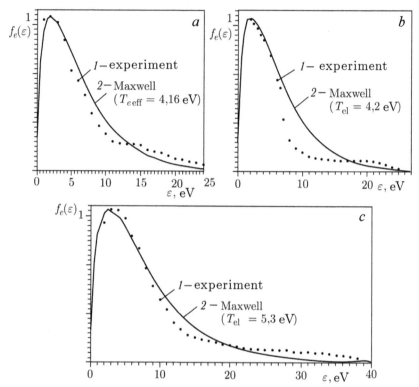

Fig. 5.23. Electron distribution functions by energy. RFCD. Air, p = 0.1 Torr, f = 1 MHz. *a)* $V_\sim$ = 550 V, *b)* $V_\sim$ = 750 V, *c)* $V_\sim$ = 850 V.

The EES of the low-pressure RFCD plasma in He and Ar have also been experimentally studied also in Refs. [119, 259].

We note general points and differences in the results obtained. We also take into account that the frequency of the RF field in our work ($f \sim 1$ MHz) was significantly lower than in [119] (27 MHz) and [259] (13.56 MHz). Because of this, for equal applied RF voltages, the plasma electron concentrations in Refs. [119, 259] were several times higher than for the present work.

We also recall that the measurements in the cited papers were made at the centre of the discharge, and in the present work in the near-electrode plasma. At the same time, the comparison of the EES plasma in the centre and the near-electrode region is valid because of the large characteristic relaxation length of electrons with respect to the energies l_e under the investigated conditions. The estimates given by the formula [111]

$$l_e = \frac{\lambda_e}{\dfrac{2m}{M} + \dfrac{v_{ee}}{v_{en}}}$$

showed that l_e exceeds the length of the discharge gap.

In the RFCD in He the received EES formed a quasi-Maxwellian group with an excess of electrons of both low and high energies (Fig. 5.20).

The EES data, presented in semilogarithmic form, reveal two groups of electrons with different effective temperatures T_{e1}, T_{e2} (Fig. 5.24). The values of the corresponding temperatures were: $T_{e1} \approx 4-6$ eV; $T_{e2} \approx 10-20$ eV. Analogous values of T_{e1} were obtained in [119, 259].

As can be seen from the EES in Fig. 5.24, with increasing RF voltage in the interval $V_\sim = 600-1000$ V, the value of T_{e1} decreases, and for $V_\sim \geq 1200$ V starts to grow. The effective electron temperature $T_{e\,\text{eff}}$, determined by the formula

$$T_{e\text{eff}} = \frac{2}{3}\langle \varepsilon_e \rangle,$$

where $\langle \varepsilon_e \rangle$ is the average electron energy obtained from the experimental distribution function $f_e(\varepsilon)$, first decreases with increasing $V_\sim$ from 600 V to 800 V, and then slowly increases in the interval $V_\sim = 800-1200$ V.

Note that in the presented energy distributions (Fig. 5.24), the energy range of the quasi-Maxwell EES region with temperature T_{e1} monotonically expands with increasing RF voltage.

An analysis of the EES data also shows that with increasing $V_\sim$ the relative number of 'superthermal' electrons (i.e., a larger number than expected for the Maxwellian function) decreases with $V_\sim \leq 1000$ V, and at $V_\sim \geq 1000$ V there is an increase.

Let us consider the possible mechanisms of formation of the received EES of the near-electrode plasma.

First of all, let us pay attention to sufficiently large values of $T_{e1} \sim 5$ eV. According to Ref. [111], this indicates that in the investigated discharge, a transition from the α-discharge to the γ-discharge was not achieved where $T_{e1} < 1$ eV was observed. Hence the conclusion follows that the main factor in creating the investigated near-electrode plasma is a rather strong RF field, concentrated at the active electrode.

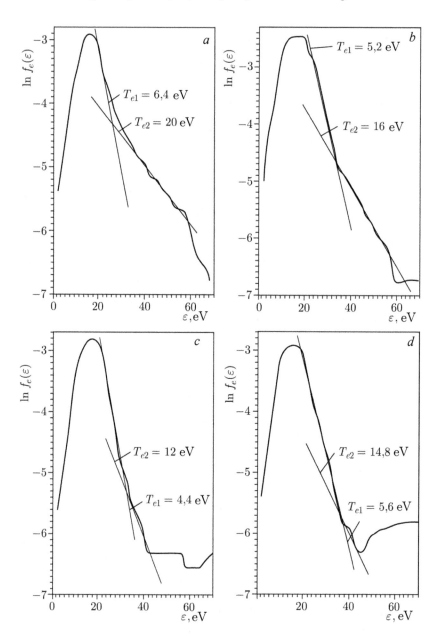

Fig. 5.24. *a, b*) The electron energy distribution functions in the semilogarithmic scale. He, $p = 0.1$ Torr, $f = 1.2$ MHz. *a*) $V_\sim = 600$ V, $T_{e\,\text{eff}} = 12.7$ eV; *b*) $V_\sim = 800$ V, $T_{e\,\text{eff}} = 11.3$ eV. *c, d*) The electron energy distribution functions in the semilogarithmic scale. He, $p = 0.1$ Torr, $f = 1.2$ MHz. *c*) $V_\sim = 1000$ V, $T_{e\,\text{eff}} = 11.7$ eV; *d*) $V_\sim = 1200$ V, $T_{e\,\text{eff}} = 11.9$ eV.

At the same time, we note another circumstance, apparently connected with the large values of the RF voltages applied in our investigations.

In [268], the measured distribution function in the RFCD plasma, maintained by an RF voltage of $V_\sim \sim 100$ V, was compared with the Maxwellian function with an effective temperature $T_{e\ \mathrm{eff}}$.

As previously reported, the experimental function $f_e(\varepsilon)$ mentioned above differed from the Maxwellian function by the excess of electrons of small and high energies.

For our conditions ($V_\sim = 600-1200$ V) the comparison of the experimental function $f_e(\varepsilon)$ with the Maxwellian function with the temperature $T_{e\ \mathrm{eff}}$ looked clearly incorrect – they differed so much (Fig. 5.24). From our point of view, this is explained by the large enrichment of the 'tail' of the function $f_e(\varepsilon)$ by high-energy electrons, which gives too high values of $T_{e\ \mathrm{eff}}$ and the corresponding Maxwellian function, which is very different from the experimental one.

Meanwhile, a comparison of the experimental functions $f_e(\varepsilon)$ with the Maxwellian ones with the temperature T_{e1} of the group of low-energy electrons gives them a relatively good correspondence (Fig. 5.20). Moreover, taking into account the measurement error and the absence of the Ramsauer effect in He, one should not attach importance to some discrepancy between the above-mentioned distributions in the low-energy region.

It makes sense to investigate in more detail the behaviour of the high-energy part of the experimental distributions $f_e(\varepsilon)$ with a change in the amplitude of the RF voltage in a sufficiently large range.

According to the data of Ref. [111], a decrease in the values of T_{e1} and $T_{e\ \mathrm{eff}}$ with an increase in $V_\sim$ signifies a significant role of the collisional processes in the plasma energy balance, and the increase of T_{e1} in the importance of collisionless processes (for example, stochastic heating of electrons by the oscillating boundary of the NESCL) electronic collisions of particles of different energy groups.

Indeed, under the conditions of [111, 119, 268] at the RFCD in He the discharge parameters, in particular, had the values: $V_\sim \sim 100$ V, $f \geq 13.56$ MHz. This provided a sufficiently effective role of collisional processes, stochastic heating of electrons in the near-electrode regions and Coulomb collisions of electrons in the discharge mechanism.

Our experimental conditions were distinguished by substantially larger values of RF voltages and relatively low field frequencies:

$V_\sim \geq 600$ V, $f = 1$ MHz. This caused the appearance in the discharge of electrons with energies $\varepsilon_e \geq 600$ eV, which led to a decrease in the intensity of collisional processes.

Since the power of the stochastic heating of the electrons is [56]:

$$P = \frac{3 f^2 V_\sim^2}{v_{Te}},$$

where v_{Te} is the average thermal velocity of electrons, then stochastic heating is expected to be weaker than in [111, 119, 268], since $P \sim f^2$.

Analogously, it was necessary to assume an appreciable manifestation of Coulomb collisions only at higher values of $V_\sim$, since the electron density is $n_e \sim f^2$ [67], and the frequencies are low.

Let us clarify the specifics of the study of the EES of the low-pressure RFCD of earlier works [111, 119, 268] and our studies.

In the papers cited, as a rule, symmetrical RFCDs with a large area of electrodes and low RF voltages ($V_\sim \sim 10^2$ V) were investigated. As a consequence, the values of the energy ε_{eb} and the density n_{eb} of the beams should be relatively low. Apparently, in connection with this in these works, the role of the beams in the formation of the EES of the plasma of the central region of RFCD is not analyzed, the possibility of the appearance of various beam instabilities is completely ignored. Although it should be done in some modes of RFCD.

The obtained results show that with increasing RF voltage, the energy interval, in which the experimental function $f_e(\varepsilon)$ approaches the Maxwellian one with the temperature T_{e1} (Fig. 5.24), widens A similar result was obtained in somewhat different physical conditions and was explained by the influence of Coulomb collisions of electrons [268].

Let us estimate the role of the Coulomb collisions in our conditions. The influence of the latter will be decisive when the collision frequency v_{ee} appreciably exceeds the characteristic energy exchange frequency of 'slow' electrons with neutral particles, that is, if the following condition is satisfied:

$$v_{ee} > v_1, \quad v_1 = \frac{2m}{M} v_{en1}.$$

The frequency of elastic collisions of electrons with a low-temperature groups with atoms is

$$V_{enl} = n_a \sigma_{en}(\varepsilon) v_{T_{el}} = P_c(\varepsilon) v_{T_{el}} = P_c(\varepsilon) \left(\frac{3kT_{el}}{m} \right)^{1/2},$$

where $P_c(\varepsilon)$ is the probability of collisions [157].

Hence, the characteristic frequency of the exchange of energy of electrons with atoms

$$V_1 = \frac{2P_c(\varepsilon)}{M} (3mkT_{el})^{1/2}.$$

The frequency of Coulomb e–e collisions for a group of 'slow' electrons [4]:

$$V_{ee} = \frac{3.7 n_e \ln \Lambda}{T_{el}^{3/2}},$$

$\ln \Lambda = 7.47 + \frac{3}{2} \lg T_{el} - \frac{1}{2} \lg n_{el}$ is the Coulomb logarithm,

The electron temperature T_{el} was determined from the slope of the semilogarithmic curves $\ln f_e(\varepsilon)$ (Fig. 5.24).

The electron density n_{el} was determined from the measurements using an energy analyzer: the electron current density in the energy analyzer j_e and the electron energy distribution function $f_e(\varepsilon)$ (Fig. 5.20):

$$j_e = \frac{1}{4} e \langle n_e v_e \rangle = \frac{1}{4} e \int_0^\infty f_e(\varepsilon) \left(\frac{2}{m} \right)^{1/2} \varepsilon^{1/2} d\varepsilon =$$

$$= \frac{e}{4} \int_0^\infty n_{el}' F_e(\varepsilon) \left(\frac{2}{m} \right)^{1/2} \varepsilon^{1/2} d\varepsilon = \frac{e}{4} n_{el}' \left(\frac{2}{m} \right)^{1/2} \int_0^\infty F_e(\varepsilon) \varepsilon^{1/2} d\varepsilon;$$

$$n_{el}' = \frac{4 j_e}{e \left(\frac{2}{m} \right)^{1/2} \int_0^\infty F_e(\varepsilon) \varepsilon^{1/2} d\varepsilon},$$

where $F_e(\varepsilon)$ is the electron energy distribution function, normalized to 1.

The $\int_0^\infty F_e(\varepsilon) \varepsilon^{1/2} d\varepsilon$ integral was calculated with the help of a computer into which the experimentally defined function $F_e(\varepsilon)$ was introduced, presented in a tabular form.

As an example, we describe the calculation of the quantities n_{el}, v_1, v_{ee} and v_{ee}/v_1 for one case: RFCD in He, $p = 0.1$ Torr, $f = 1.2$ MHz, $V_\sim = 1000$ V, $\int_0^\infty F_e(\varepsilon) \varepsilon^{1/2} d\varepsilon = 4.22$ eV$^{1/2}$.

$$n'_{el} = \frac{4 \cdot 0.03 \cdot 3 \cdot 10^9 \cdot (9.1 \cdot 10^{-28})^{1/2}}{4.8 \cdot 10^{-10} \cdot 1.41 \cdot 4.22 \cdot (1.6 \cdot 10^{-12})^{1/2}} = 3 \cdot 10^9 \text{ cm}^{-3}.$$

The collector of the energy analyzer received a reduced flux of electrons due to incomplete transparency of the screening and control grids of the energy analyzer. The permeability of each grid was 75%. Thus, the electron flux entered the collector 4 times weakened. Therefore, the electron density in the plasma n_e should be higher by the corresponding number of times than that recorded by the energy analyzer

$$n_{el} = 4n'_{el} = 4 \cdot 3 \cdot 10^9 = 1.2 \cdot 10^{10} \text{ cm}^{-3}.$$

Further we have

$$v_1 = \frac{2 \cdot 0.1 \cdot 16}{6.64 \cdot 10^{-24}} \left(3 \cdot 9.1 \cdot 10^{-28} \cdot 4.5 \cdot 1.6 \cdot 10^{-12} \right)^{1/2} = 0.8 \cdot 10^5 \text{ s}^{-1}.$$

$$\ln \Lambda = 9.53, \quad v_{ee} = \frac{3.7 \cdot 1.2 \cdot 10^{10} \cdot 9.53}{\left(4.5 \cdot 1.16 \cdot 10^4 \right)^{3/2}} = 0.4 \cdot 10^5 \text{ s}^{-1};$$

$$\frac{v_{ee}}{v_1} = \frac{0.4 \cdot 10^5}{0.8 \cdot 10^5} = 0.5.$$

The results of similar calculations for different values of RF voltage are given in Table. 5.1.

The data presented show that with an increase in $V_\sim$ the ratio v_e/v_1 increases, and a slight decrease in this quantity at $V_\sim = 1200$ V is explained by the increase in the heating of the electrons, the mechanism of which is discussed below.

Since at $V_\sim \geq 1000$ V the frequency v_{ee} becomes close to the characteristic frequency v_1, the observed tendency of the distribution function $f_e(e)$ to the Maxwellian frequency with an increase in $V_\sim$ finds a satisfactory explanation by the action of the Coulomb collisions. The maximum deviation from the Maxwellian distribution

Table 5.1. He, $p = 0.5$ Torr, $f = 1.2$ MHz, RFCD with symmetrical electrodes

$V_\sim$, B	j_e,mA/cm^2	T_{el},eV	v_1, 10^{-5}s^{-1}	n_{el},cm^{-3}	v_{ee}, 10^{-5}s^{-1}	v_{ee}/v_1
600	13	6.5	0.82	$5.2 \cdot 10^9$	0.09	0.1
800	19	5.2	0.77	$8 \cdot 10^9$	0.20	0.3
1000	30	4,5	0.80	$1.2 \cdot 10^{10}$	0.36	0.5
1200	50	5.6	0.76	$2 \cdot 10^{10}$	0.32	0.4

is observed at higher electron energies, where the Coulomb collision frequency is small.

Let us now investigate the observed enrichment of the function $f_e(e)$ by non-equilibrium electrons of increased energies. In this case, let us pay attention to the fact that when the amplitude of the RF voltage increases monotonously, the excess of non-equilibrium electrons first decreases and then begins to increase.

The mechanism of the formation of the EES of the plasma at the grounded electrode is clearly non-local because of the high thermal conductivity of the plasma and the long relaxation times of the near-electrode electron beam (NEEB) in energy. Therefore, the investigated distribution function $f_e(\varepsilon)$ forms the entire set of the processes listed above for both electrodes.

Analyzing the physical processes in the plasma of the RFCD under consideration, it should be taken into account that, depending on the amplitude of the RF voltage and the phase of the RF field period, the mean free paths of the beam electrons for the various elementary processes (elastic collisions, excitation and ionization of atoms) can be either smaller or substantially exceed the length of the discharge gap. This is due to the previously mentioned modulation of the electron energy of the beams during the RF field period. Thus, the NEEB should produce a continuous spatial scanning of the interelectrode gap along the axis of the discharge tube. Let us consider specific values of the mean free paths of beam electrons for RFCD in He at $p = 0.1$ Torr and $d = 6$ cm.

As an example, we estimate the mean free path of electrons with energies $e_e = 36$ eV and 60 eV in a gas, using the data of [4]. We have:

$$\varepsilon_e = 36\,\text{eV}, \quad P_c(36\,\text{eV}) = 0.7\,\text{cm}^{-1}; \quad \lambda_e = 1/P_c = 1.43\,\text{cm};$$

$$\varepsilon_e = 60\,\text{eV}, \quad P_c(60\,\text{eV}) = 0.25\,\text{cm}^{-1}; \quad \lambda_e = 1/P_c = 4.0\,\text{cm}.$$

Obviously, with a further increase in ε_e, we obtain $\lambda_e > d$.

Let us now estimate the ionization length of the mean free path λ_i of electron beams. We have:

$\varepsilon_e = 200$ eV; $P_i = 1.25$ cm^{-1} · Torr^{-1} – the probability of ionization collision under normal conditions [4].

$$\lambda_i\,(200\,\text{eV}) = \frac{1}{P_i} = \frac{1}{0.1 \cdot 1.25} = 8\,\text{cm} > d.$$

$$\varepsilon_e = 500\ \text{eV}, \quad P_i(\varepsilon) = 3.54 \cdot 10^{16} \cdot \sigma_i(\varepsilon) p.$$

We find the ionization cross section $\sigma_i(\varepsilon)$ by Thomson's formula [4]:

$$\sigma_i(\varepsilon) = \frac{\pi e^4}{I} \cdot \frac{\varepsilon - I}{\varepsilon^2},$$

where I is the ionization potential He, $I = 24.5$ eV,

$$\sigma_i(\varepsilon) = \frac{3.14 \cdot 4.8^4 \cdot 10^{-40}}{24,5 \cdot 1,6 \cdot 10^{-12}} \cdot \frac{(500 - 24.5) \cdot 1.6 \cdot 10^{-12}}{(500 \cdot 1,6 \cdot 10^{-12})^2} = 5.1 \cdot 10^{-18} \text{ cm}^2.$$

$$P_i(500 \text{ eV}) = 3.54 \cdot 10^{16} \cdot 5.1 \cdot 10^{-18} \cdot 0.1 = 1.81 \cdot 10^{-2} \text{ cm}^{-1}.$$

$$\lambda_i = \frac{1}{P_i} = \frac{1}{1.81 \cdot 10^{-2}} = 55 \text{ cm}, \quad \lambda_i \gg d.$$

The data presented indicate that the time-averaged characteristic free path lengths of electron beams substantially exceed the length of the discharge gap.

The situation should change qualitatively, when in the conditions in the discharge are fabourable for the excitation of beam instabilities by electron beams which sharply reduce the relaxation length of the beams by momentum and energy, which will be considered below.

The analysis of the obtained distribution functions $f_e(\varepsilon)$ (Fig. 5.20) and their representation in the semilogarithmic form $\ln f_e(\varepsilon)$ (Fig. 5.24) leads to the following conclusions regarding the mechanism of their formation.

There is the reason to believe that before reaching the excitation threshold (at $V_\sim \geq 1$ kV), the form of the EES of the near-electrode plasma is determined mainly by the heating of the electrons by the RF field in the vicinity of the active electrode, and the high-energy part of the EES is enriched by the scattering of beams from both electrodes in collisions with atoms.

This is confirmed by the fairly large values of the temperature T_{e1} of 'slow' electrons, close to those in the plasma of the positive column of DCGD, when the electrons are in thermodynamic equilibrium with the electric field in the plasma. The observed decrease in T_{e1} with increasing $V_\sim$ indicates the collisional mechanism of electron heating.

The observed decrease in the relative number of fast electrons in the EES (Figs. 5.20, 5.24) in the voltage range $V_\sim = 600-1000$ V can be explained by the increase in the energies of the electron beams ε_{eb} and the corresponding increase in their mean free path, which decreases the scattering of beams by atoms.

When the amplitude of the RF voltage is reached, $V_\sim \geq 1200$ V, the decrease in temperature of T_{e1} stops and its growth begins. At the same time, an increase in the relative number of electrons in the high-energy part of the EES is observed (Figs. 5.20, 5.24). An explanation of this is given below.

Let us consider the obtained electron distribution functions $f_e(\varepsilon)$ in the near-electrode plasma near the grounded electrode in RFCD in Xe (Figs. 5.21, 5.22). The characteristics of the physical conditions in the near-electrode plasma for the two frequencies of the RF field of $f = 0.6$ MHz and 1 MHz are presented in Table 5.2 and 5.3 respectively. At the same time, calculations similar to those for the RFCD in He w.

We discuss the general properties of the obtained distributions of $f_e(\varepsilon)$ for RFCD in Xe. At low RF voltages ($V_\sim < 1000$ V), the distribution function of low-energy electrons is close to the Druvesteyen distribution. The greatest enrichment of the function $f_e(e)$ by high-energy electrons was observed in the RFCD at a frequency $f = 1$ MHz at a value $V_\sim = 600$ V. Under these conditions, the distribution of $f_e(e)$ consists of two groups of electrons.

As can be seen from the presented distributions of $f_e(\varepsilon)$, with a monotonic increase in $V_\sim$, a sharp decrease in the number of high-energy electrons occurs, a smooth transition of the function $f_e(\varepsilon)$ to a single-group distribution, and the transformation of the Druvesteyen distribution to Maxwellian under certain physical conditions.

Table 5.2. The plasma parameters of RFCD in Xe, $p = 0.1$ Torr, $f = 0.6$ MHz

Options	$V\sim$, V			
	800	1000	1200	1400
n_e, 10^{-9}cm^{-3}	1	2	5	8
T_{e1}, eV	-	-	2.2	-
$T_{e\ eff}$, eV	3.6	3.7	3.3	3.4
ω_{e0}, 10^{-8}s^{-1}	19	26	40	51
ν_{en}, 10^{-8}s^{1}	14	15	5.2	14
$\delta\nu_{en}$, 10^{-3}s^{-1}	11	12	4.3	11
ν_{ee}, 10^{-3}s^{-1}	4.3	8.1	42	51
δ_1, 10^{-8}s^{-1}	0.7	1.1	3.5	4.3
δ_2, 10^{-8}s^{-1}	1.5	2.1	3.2	4.1
δ_1/ν_{en}	0.05	0.07	0.67	0.31
δ_2/ν_{en}	0.11	0.14	0.62	0.29
L_1/d	4	2.8	0.95	0.88

Table 5.3. The plasma parameters of RFCD in Xe, p = 0.1 Torr, f = 1 MHz

Options	$V_{\sim}$, V			
	600	800	1000	1200
n_e, 10^{-9}cm^{-3}	1.5	2.5	3.5	9
T_{e1},eV	–	–	1.6	2.2
T_{eeff},eV	4.3	3.9	3.5	3.0
ω_{e0}, 10^{-8}s^{-1}	22	28	33	54
ν_{en}, 10^{-8}s^{-1}	13	11	10.5	7.2
$\delta\nu_{en}$, 10^{-3}s^{-1}	10	9	8.2	5.9
ν_{ee}, 10^{-3}s^{-1}	4.9	9.5	15	35
δ_1, 10^{-8}s^{-1}	0.93	1.4	1.9	4.7
δ_2,10^{-8}c^{-1}	1,8	2.2	2.7	4.3
δ_1/ν_{en}	0.07	0.13	0.18	0.65
δ_2/ν_{en}	0.14	0.20	0.25	0.60
L_1/d	2.7	2.0	1.7	0.7

We note that Xe is a gas with an essentially pronounced Ramsauer effect.

Earlier investigations of the RFCD in another Ramsauer gas – Ar – showed that the distribution function $f_e(\varepsilon)$ in this case corresponds to the Druvesteyen distribution [111].

A comparison of the obtained functions $f_e(\varepsilon)$ for RFCD in Xe and Ar shows their good qualitative similarity (Fig. 5.25).

The broader peak of the function $f_e(\varepsilon)$ in the low-energy region in Xe, in comparison with that observed in Ar, is explained by the difference in the quantitative proportions during the cross section of collisions of electrons in the vicinity of the Ramsauer minimum for these gases.

The data presented in Tables 5.2 and 5.3 show that in certain RFCD modes the following condition is fulfilled

$$\delta\nu_{en} < \nu_{ee},$$

and the contribution of Coulomb collisions exceeds the influence of elastic collisions of electrons with atoms on the process of formation of the distribution of $f_e(\varepsilon)$, acquiring the character of the Maxwellian distribution (Figs. 5.21, 5.22).

The mechanism of enrichment of the EES of the plasma by high-energy electrons under these conditions is considered below.

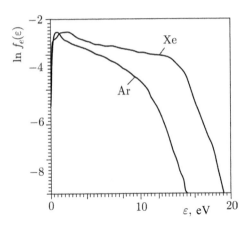

Fig. 5.25. Functions of electron energy distribution in the semilogarithmic scale in argon and xenon. RFCD. $p = 0.1$ Torr, $f = 1$ MHz, $V_{\sim} = 600$ V.

Table 5.4. Parameters of plasma RFCD in air, $p = 0.1$ Torr, $f = 1$ MHz

Options	$V_{\sim}$, V					
	550		750		850	
n_e, 10^{-9} cm^{-3}	1.0	10	2.5	25	4.0	40
T_{e1}, eV	4.1		4.2		5.3	
$T_{e\,eff}$, eV	4.2		5.3		–	
ω_{e0}, 10^{-8}s^{-1}	18	57	28	90	36	110
ν_{en}, 10^{-8}s^{-1}	5.3		4.9		5.3	
$\delta\nu_{en}$, 10^{-3}s^{-1}	1100		980		1100	
ν_{ee}, 10^{-3}s^{-1}	3.6	36	8.5	85	9.6	96
δ_1, 10^{-8}s^{-1}	1.1	6.2	2.1	12	3.0	16
δ_2, 10^{-8}s^{-1}	1.4	4,5	2.2	7.2	2.9	8.7
δ_1/ν_{en}	0.21	1.17	0.43	2.45	0.51	3.00
δ_2/ν_{en}	0.25	0.85	0.45	1.47	0.55	1.65
L_1/d	2.12	0.38	1.27	0.22	0.95	0.18

Let us consider the results of an experimental study of the EES of the near-electrode plasma of the RFCD in air (Fig. 5.23), as well as data of the analysis of physical conditions in the investigated discharge, given in Table 5.4.

Air, in contrast to atomic gases (He, Xe), is a mixture of molecular gases (N_2, O_2) having additional internal degrees of freedom of energy distribution – vibrational and rotational. As a result, the space-time relaxation of the near-electrode electron beams with respect to energy should occur much faster than in atomic gases. The average fraction of the energy transferred by electrons in collisions with air molecules is $\delta = 2 \cdot 10^{-3}$ [4]. Therefore, in the RFCD in the air, the condition $\delta v_{en} \gg v_{ee}$ is always fulfilled. So the influence of e–e Coulomb collisions on the formation of the function $f_e(\varepsilon)$ could be neglected. As can be seen from the obtained distributions $f_e(e)$ (Fig. 5.23), the latter were significantly different from the Maxwellian distribution.

The maximum discrepancies in the distribution $f_e(\varepsilon)$ with Maxwellian were observed in the energy range $\varepsilon_e = 6$–16 eV (Fig. 5.23 *b*). It is this interval in which the energy thresholds of ionization, dissociation, and other inelastic collisions of electrons with N_2, O_2 molecules occur.

An additional discussion of the mechanism for the formation of the EES of the RFCD plasma in air is given below.

5.5. EES of the plasma of asymmetric RFCD

Particular attention was paid to the experimental study of the physical properties of the plasma near an earthed electrode in an asymmetric RFCD.

In accordance with the existing ideas, this plasma should be created by a dense high-energy electron beam from the active electrode of a small area. In the vicinity of the grounded electrode, it is estimated that the conditions for the effective energy deposition of the beam in the plasma are most probable.

The nature of the EES of such plasma, its formation mechanisms, and also the study of the possibilities of controlling the electron energy distribution function in such a plasma medium are interesting.

Thanks to the effective methodological method used – the use of a highly asymmetric RFCD, it was possible to significantly increase the density of electron beams.

As a result, it became possible, with the help of an energy analyzer, to examine the entire high-energy part of the EES of near-electrode plasma. The latter was not possible with analogous measurements in the symmetrical RFCD, as well as in measurements by other authors in the central region of the RF discharge by the probe method, in principle unsuitable for studying the high-energy region of EES [111, 119].

The problem of registering the initial section of the power plant with the help of an energy analyzer remains in this case. However, the main task of the present work was to study the properties of anomalously wide EES and, in particular, their high-energy regions.

The experimental data presented here refer to measurements of quasi-stationary quantities. At the same time, the delay curves were obtained by measuring with the V7-27A voltmeter the voltage on the load resistance of the energy analyzer.

Note, however, that most of the measurements were carried out by means of time-resolved energy analyzer signals. The corresponding results are given below.

Let us consider the most characteristic data on the study of quasi-stationary EESs of near-electrode plasma of an asymmetric RFCD.

The EES of the near-electrode plasma of the asymmetric RFCD (ARFCD) at $p = 0.1$ Torr, through the computer differentiation of the delay curve from the collector of the energy analyzer $I_c(V_c)$, is shown in Fig. 5.26.

Another example of an EES made in RFCD at a higher RF voltage is shown in Fig. 5.27.

The results of an experimental study of the dependence of the high-energy part of the EES of the ARFCD plasma on the voltage amplitude $V_\sim$ at $p = 0.07$ Torr are shown in Fig. 5.28.

A number of features of the above EES should be noted.

In the case of ARFCD in air at a pressure $p = 0.1$ Torr and $V_\sim = 600$ V, the high-energy part of the EES will have a very wide plateau-shaped region, relatively weakly modulated in amplitude, at the end of which there is a high quasi-monoenergetic peak separated by a wide dip from the main region of EES.

We note that the maximum values of the electron energies $\varepsilon_{e\,max}$, recorded in the measurement of the EES of the ARFCD plasma, are close to the values $\varepsilon_{e\,max} \approx 2eV_\sim$ theoretically predicted in Ref. [78].

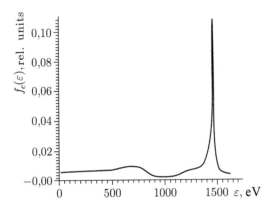

Fig. 5.26. The energy distribution function of electrons. ARFCD. Air, $p = 0.1$ Torr, $f = 1$ MHz, $V_\sim = 600$ V.

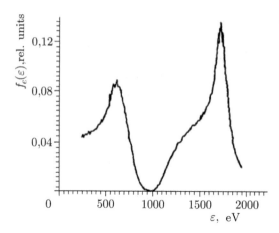

Fig. 5.27. The energy distribution function of electrons. ARFCD. Air, $p = 0.1$ Torr, $f = 1$ MHz, $V_\sim = 750$ V.

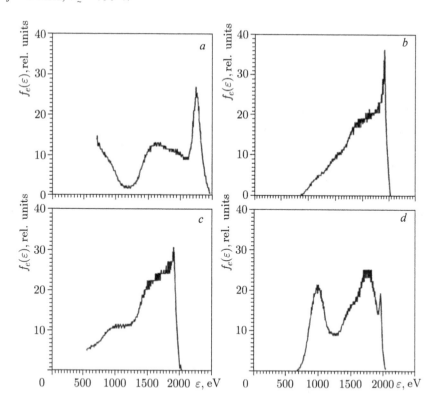

Fig. 5.28. Dependence of the high-energy part of the electron distribution of the near-electrode plasma with respect to energy on RF voltage. ARFCD. Air, $p = 0.07$ Torr, $f = 1$ MHz. *a)* $V_\sim = 850$ V, *b)* $V_\sim = 900$ V, *c)* $V_\sim = 950$ V, *d)* $V_\sim = 1000$ V.

Indeed, for the case shown in Fig. 5.28 we get:

$$\varepsilon_{e\,\text{max}} = 2eV_{\sim} \approx 2100 \text{ eV},$$

which corresponds exactly to the maximum values of the electron energies recorded in the experiment.

Comparing the EES in Figs. 5.26 and 5.27, we see that with an increase in the amplitude $V_{\sim}$, the beam peak of the EES is noticeably blurred with respect to energy.

An essentially different picture of the behaviour of the high-energy part of the EES is observed in the case of ARFCD with voltages $V_{\sim} \geq 850$ V and even lower gas pressure $p = 0.07$ Torr (Fig. 5.28). At the same time, the maximum at the end of the EES, while remaining in the neighbourhood of the value of the energy $\varepsilon_{e\,\text{max}}$, is no longer isolated from the rest of the spectrum, but closes with it, and with increasing voltage $V_{\sim}$ the final peak of the EES is gradually degrading, tending to attenuation. The central region of the EES with a monotonic increase in $V_{\sim}$ is essentially transformed.

An explanation of the behaviour of the EES of the ARFCD plasma described above is given below.

We shall determine, using energy analyzer measurements, the concentration of high-energy electrons n_{eh} in the ARFCD plasma.

We use for this purpose the experimental curve of the EES of the ARFCD plasma in air (Fig. 5.27). From the oscillogram of the signal $I_c(t)$ and the known area of the entrance opening S_a, we find the current density $j_e = I_e/S_a$ on the collector of the energy analyzer. For the case under consideration, $j_e = 300$ mA · cm². The current density j_e is:

$$j_e = e\langle n_e v_e \rangle = e \int_{\varepsilon_{el}}^{\infty} f_e(\varepsilon) v_e \, d\varepsilon,$$

where $f_e(\varepsilon) = n_{\text{eh}} F_e(\varepsilon)$ with the normalization conditions:

$$\int_{n_{el}}^{\infty} F_e(\varepsilon) d\varepsilon = 1, \quad v_e = \left(\frac{2\varepsilon_e}{m}\right)^{1/2}, \quad \varepsilon_{el} \geq 50 \text{ eV}.$$

From this we get

$$j_e = e \int_{\varepsilon_{el}}^{\infty} n_{\text{eh}} F_e(\varepsilon) \left(\frac{2\varepsilon_e}{m}\right)^{1/2} d\varepsilon = en_{\text{eh}} \left(\frac{2}{m}\right)^{1/2} \int_{\varepsilon_{el}}^{\infty} F_e(\varepsilon) \varepsilon_e^{1/2} d\varepsilon$$

and then

$$n_{eh} = \frac{j_e}{e} \cdot \left(\frac{m}{2}\right)^{1/2} \cdot \left(\int_{\varepsilon_{el}}^{\infty} F_e(\varepsilon)\varepsilon^{1/2}d\varepsilon\right)^{-1}.$$

Entering the experimental function $F_e(\varepsilon)$ into the computer in tabular form, we obtain the values of the integral $\int_{n_{el}}^{\infty} F_e(\varepsilon)\varepsilon_e^{1/2}d\varepsilon$.

For the experimental curve of the EES in Fig. 5.27, we obtain:

$$\int_{n_{el}}^{\infty} F_e(\varepsilon)\varepsilon_e^{1/2}d\varepsilon = 0.402 \ \mathrm{erg}^{3/2}.$$

Thus, for the considered case

$$n_{eh} = \frac{300 \cdot 10^{-3} \cdot 3 \cdot 10^9}{4.8 \cdot 10^{-10} \cdot \sqrt{2} \cdot 0.402} \cdot (9.1 \cdot 10^{-28})^{1/2} = 10^5 \ \mathrm{cm}^{-3}.$$

Thus, the density of electrons with energies in the range $50 \leq \varepsilon_e \leq 2100$ eV, in the investigated plasma is $n_{eh} = 10^5$ cm^{-3}. The absolute value of n_{eh} obtained is discussed below.

5.6. The time course of physical processes in the RFCD and its effect on the electronic energy spectrum of near-electrode plasma

The time dependence of the RFCD characteristics is an obvious inherent property of it. In this case, we are talking about the time course of physical processes within a single period of the RF field.

As is well known, the characteristic times for the variation of the plasma parameters n_e, $f_e(e)$, and the density of the excited atoms n_k^* vary significantly in magnitude.

Different times of relaxation characterize the individual parts of the EES of the plasma, especially occupying a wide energy interval. Thus, for example, the relaxation time of the most numerous group of 'slow' electrons considerably exceeds the corresponding time for the high-energy 'tail' of the energy distribution.

Hence, it is clear that the study of the time dependence of the EES of plasma is important for studying the physical mechanism of formation of the latter.

The experimental technique for studying the time dependence of the EES of the plasma was the study of the fluxes of charged particles to a grounded electrode with an energy analyzer used in various versions in the measuring circuit.

In one case, the control grid was connected to the collector, and the work was done in the 'shielded probe' mode. At the same time, no analyzer voltages were used in the device, the plasma flow through the entrance hole, tightened by a grid, entered the collector, from where the charge passed through the resistance of the load to the ground. The resistance signal R_{load} was investigated with the help of a broadband double-beam oscilloscope S1-74. On one channel of the oscilloscope, the total current of electrons and ions was observed. However, due to the essentially different time course of the electron and ion currents, it was possible to observe confidently the dynamics of both currents during the period of the RF field. A signal proportional to the RF voltage was applied to the other channel to determine the phase ratio of the RF field period and the investigated dynamics of the charged particle fluxes.

The second regime was used to obtain plasma EES during a certain phase of the RF period, or to study the time course of the electron flux having a given energy range during the RF field period. In this mode, there were two options for supplying the electron analyzing voltage: a) to the analyzing grid and b) to the collector. In experiments, preference was often given to the latter variant due to better repeatability of the measurement results.

The undefined shape of the observed signals was provided by the optimal choice of load resistance R_{load} = 560 Ohm and capacitance in the measuring circuit C_m = 20 pF. Hence the time constant of the measuring circuit was $\tau = R_{load} \, C_m = 560 \cdot 20 \cdot 10^{-12} = 10^{-8}$ s, which is two orders of magnitude shorter than the duration of the RF field period. The bandwidth of the oscilloscope was 50 MHz.

The form of the oscillograms of the fluxes of charged particles on the grounded electrode was given earlier. An example of graphical processing of oscillograms $I_c(t)$ in the absence of analyzing voltages is shown in Fig. 5.29.

The obtained oscillograms $I_c(t)$ make it possible to obtain the following information: 1) instantaneous values of the total current of charges entering the analyzer $I_c(t)$; 2) the maximum values of ionic I_i and electronic I_e currents in the analyzer; 3) the amplitude of the variable component of the ion current ΔI_i; 4) the half-width of the duration of the electron pulse from the plasma τ_e; 5) the phase of the maximum of the electron pulse; 6) certain information about the dynamics of the NESCL consists in the asymmetry of the edges of

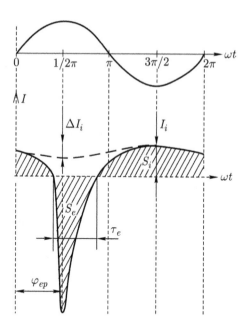

Fig. 5.29. Scheme of the analysis of the total current on the electrode during one period of the RF field.

the electron pulse; in particular, the trailing edge of this pulse is a natural delay curve of the plasma electrons by an increasing potential barrier of the NESCL and, in principle, allows us to determine the EES.

The method for obtaining time-resolved plasma EESs is based on the characteristic form of the oscillogram $I_c(t)$ (Figs. 2.24 and 2.25), where the sharp maximum of the electron pulse is observed. The position of the latter corresponds to the phase of the RF field period, close to the middle of the positive half-period. Gradually locking the electronic pulse by the analyzing voltage, it is possible to obtain a delay curve and then the electron energy spectrum corresponding to a certain moment of the RF field period. The last assertion is valid if the relaxation time of the EES under investigation is much less than the period of the RF field, which under the investigated conditions took place.

The EESs obtained in this way were of the greatest interest. The point is that, as our investigations have shown, it is in the vicinity of this phase that the electron beam energy is maximal, and the potential barrier for electrons before the grounded electrode is minimal. Thus, for the phase $\varphi = 3\pi/2$, the electron energy spectrum will be the widest for the entire period of the RF field and its perturbation by the near-electrode potential barrier will be minimal.

In the course of the work it was shown that obtaining oscillograms of charged particle fluxes to the electrode at various delay voltages in the energy analyzer can be not only a method for determining EES but also a methodical tool for studying the time course of physical processes in the plasma under investigation. Let's show it on examples.

The obtained oscillogram of the flows of charged particles on the grounded electrode in the asymmetric RFCD is shown in Fig. 5.30.

The previously presented oscillograms of fluxes of charged particles on a grounded electrode in a symmetrical RFCD (Figs. 2.24, 2.25) differ significantly in appearance from the oscillogram in the case of asymmetric RFCD (Fig. 5.30).

In the asymmetrical RFCD, the electronic part of the oscillogram consists of two consecutive pulses 1 and 2 (Fig. 5.30). By applying various inhibitory voltages it was established that the electrons of the second pulse have energies much greater than those of the electrons of the first pulse. The electrons of the second pulse were not locked even by a voltage $V_k = -620$ V.

An explanation of these results will be given below.

The electronic part of the oscillogram $I_c(t)$ in the case of a symmetrical RFCD consisted of a single pulse at $V_c = 0$. However, when the blocking voltages V_c were applied, the nature of the electron pulse changed significantly (Fig. 2.38). It can be seen from these oscillograms that with a monotonic increase in V_c, the sharp tip of the electron pulse becomes flatter and shifts toward later instants of time. This behaviour of the electron pulse can be explained by

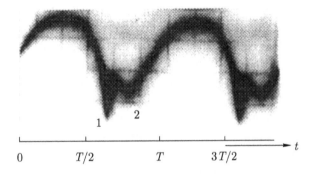

Fig. 5.30. The oscillogram of the signal of the energy analyzer $V_c(t)$. Asymmetrical RFCD. Air, $p = 0.1$ Torr, $f = 1.2$ MHz, $V_- = 1750$ V.

the fact that the electron pulse observed at $V_c = 0$ is the sum of two pulses, where the amplitude of the first pulse is dominant (Fig. 2.38).

With an increase in the negative voltage V_c, the amplitude of the first pulse decreases, while the second remains unchanged. For a sufficiently large value of $V_c \geq -50$ V the first pulse is practically suppressed and only the second pulse remains, as a result of which the vertex of the electron pulse shifts towards later instants of time (Fig. 2.38). In this case, as in ARFCD, the electrons of the second pulse have energies substantially greater than those of the electrons of the first pulse.

Hence it follows that the application of the blocking voltages V_c makes it possible to detect the impulse from the electron beam of the active electrode, which is masked, weaker than in the ARFCD.

Additional possibilities for clarifying the physical meaning of the oscillograms $I_c(t)$ give an application to the discharge gap of a constant magnetic field $\mathbf{H}_0$, perpendicular to the axis of the discharge tube, when the second pulses of the oscillogram practically disappeared. The latter gives grounds for assuming that the second pulses are due to directed fluxes, and the first pulses are caused by chaotically moving electrons.

Let us consider the experimental data obtained, characterizing the time course of physical processes in the near-electrode plasma.

First of all, let us dwell on the time dependence of the electron fluxes on the electrode for various discharge regimes and braking voltages V_c in the energy analyzer.

The analysis of oscillograms $I_c(t)$ for a symmetric RFCD is given earlier.

A similar situation was observed on the oscillograms obtained in the case of ARFCD at an air pressure $p = 0.1$ Torr (Fig. 5.30).

It should be noted here that quantitatively the shape of the oscillograms $I_c(t)$ for ARFCD depends strongly on the gas pressure. Thus, with a slight decrease in pressure from 0.1 to 0.07 Torr, an abrupt growth of the amplitude of the first electron pulse is observed on the oscillogram. This fact indicates that there is some kind of threshold physical process that causes intensive generation of 'hot' electrons in the near-electrode plasma at low pressure.

The proposed physical mechanism explaining the time dependence of the oscillograms $I_c(t)$, with the corresponding quantitative estimates, will be given below.

Let us now turn to the experimental time-resolved EES of the near-electrode plasma of the investigated RFCD.

Studies were carried out to compare the EES of a symmetric RFCD plasma, obtained quasistationally and in the time resolution mode (Fig. 5.31).

It can be seen that the EES obtained with a time resolution is shifted to the region of higher energies, in comparison with the quasi-stationary one. This was to be expected, taking into account the previously described procedure for obtaining an EES corresponding to the phase $\varphi = 3\pi/2$.

A typical time-resolved ($\varphi \approx 3\pi/2$) general form of the EES of the ARFCD plasma is shown in Fig. 5.32. Fragments of this EES on the energy axis are shown in Fig. 5.32 *a, b, c*.

A set of EESs for the near-electrode plasma of the ARFCD in air at $p = 0.07$ Torr for different values of the RF voltage amplitude is shown in Fig. 5.33.

Comparison of the resolved in time EES of the ARFCD plasma with quasi-stationary ones reveals their similarity in general terms.

However, the experimental studies carried out have also shown that the EES of the near-electrode plasma of the ARFCD changes during the period of the RF field. This is confirmed by the difference between the EES, averaged over many periods of the RF field and the time-resolved EES, measured for the phase $\varphi = 3\pi/2$. In the first

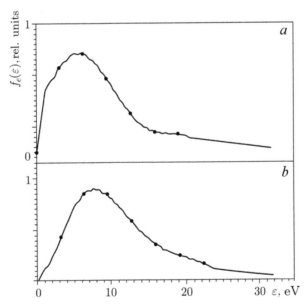

Fig. 5.31. Comparison of the quasi-stationary and time-resolved electron distribution functions with respect to energies. RFCD. Air, $p = 0.2$ Torr, $f = 1$ MHz, $V_\sim = 450$ V. *a*) quasi-stationary, *b*) for the phase of the field period $3\pi/2$.

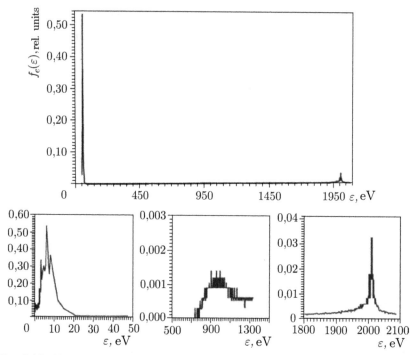

Fig. 5.32. The type of the full EES of the near-electrode plasma of ARFCD and its individual energy intervals. Air, p = 0.1 Torr, f = 1.2 MHz, V = 850 V. *a)* $0 < \varepsilon_e < 50$ eV; *b)* $500 < \varepsilon_e < 1500$ eV; *c)* $1800 < \varepsilon_e < 2100$ eV.

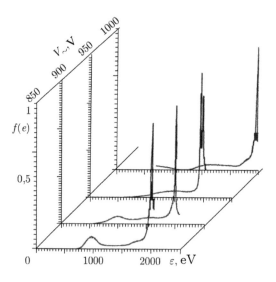

Fig. 5.33. The electron energy distribution functions corresponding to the phase of the RF field $\varphi = 3\pi/2$ (the electrons of the 'tail' of the distribution) for different RF voltages. RFCD. Air, p = 0.07 Torr, f = 1 MHz.

case, the beam peak of the EES is somewhat blurred in energy, and in the second it is close to a monoenergetic one.

The main factors responsible for a significant change in the EES of the near-electrode plasma near the grounded electrode during the RF field are the near-electrode beam modulated by density and energy from the active electrode and collapsing during each period of the RF field of the NESCL.

More details of these factors are discussed below.

5.7. Influence of the transverse magnetic field on the time course of the EES of the near-electrode plasma of ARFCD

When studying the time course of electron fluxes to the RFCD electrode, a fairly weak uniform transverse magnetic field $0 \leq H_0 \leq 60$ Oe was used.

Measurements with a magnetic field were made in ARFCD in air with typical discharge parameters. In this case, the value of the effective RF voltage was $V_\sim = 1750$ V. Thus, the maximum value of the beam electron energy from the active electrode was $\varepsilon_{e\ max} \approx 5$ keV.

As was reported above, in the absence of a magnetic field in this ARFCD, the oscillogram of the current to the electrode contained two electronic pulses (Fig. 5.30).

The curves in Fig. 5.34 of the oscillograms $I_c(t)$ show that the pulse 2 of the beam electrons was not locked by the braking voltages $V_c \leq 1$ kV. The pulse 1 corresponding to the 'hot' electrons for the considered ARFCD was locked at a voltage $V_c \leq 620$ V.

When the magnetic field was applied, an anomalous heating of the electrons of the near-electrode plasma was observed even for a weak magnetic field $H_0 \sim 20$ Oe (Fig. 5.34).

The oscillograms obtained show that pulse 1, which is almost locked by $V_c = -620$ V in the absence of a magnetic field, increases sharply when the magnetic field $H_0 = 20$ Oe is switched on. As can be seen from the oscillograms, when the magnetic field intensity is monotonously increased, the amount of 'hot' thermal electrons in the plasma rapidly increases. In this case, the pulse 2 of the beam electrons decreased and was due to scattered electrons.

Just as in the absence of a magnetic field, on the oscillogram (Fig. 5.34), the pulse 1 of 'hot' electrons soon ceases, followed by a pulse 2.

Concerning the detected effect of an increase in the number of high-energy chaotic electrons of the RFCD plasma under the action

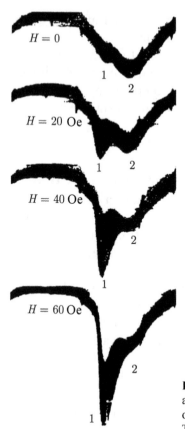

$H = 0$

1

2

$H = 20$ Oe

1   2

$H = 40$ Oe

2

1

$H = 60$ Oe

2

1

Fig. 5.34. Signals from the collector of the energy analyzer of charged particles for different values of the transverse magnetic field. Air, $p = 0.07$ Torr, $f = 1.2$ MHz, $V_{\sim} = 1750$ V, $V_{back} = -620$ V.

of a magnetic field, it can be suggested that this is due to the action of two mechanisms.

The first of these mechanisms is due to the magnetization of the plasma electrons, their retardation from the discharge, which leads to an increase in the electron density n_e. It is shown below that this creates more favorable conditions for the energy relaxation of the electron beam in the plasma, leading to the enrichment of the latter by the 'hot' electrons.

Let us estimate for these conditions the value of the Hall parameter h characterizing the degree of plasma magnetization: $h = \omega_{ec}\tau_{en}$, where $\omega_{ec} = eH/mc$ is the electron cyclotron frequency, and $\tau_{en} = 1/v_{en}$ is the time between collisions of an electron with neutral particles.

For RFCD in air at $p = 0.1$ Torr, $T_{e\,eff} \approx 0.2$ eV,

$$v_{T_e} \approx 8 \cdot 10^7 \text{ m/s}, \quad H_0 = 40 \text{ Oe}, \quad P_c = 20 \text{ cm}^{-1} \text{ [157]},$$

$$\omega_{ec} = 7 \cdot 10^8 \text{ s}^{-1}, \quad \tau_{en} = 0.8 \cdot 10^{-8} \text{ s}; \quad h = 7 \cdot 10^8 \cdot 0.8 \cdot 10^{-8} \approx 5.6.$$

Thus, the bulk of the slow electrons is magnetized to a sufficient degree.

The second electron heating mechanism can be associated with the excitation in the magnetic field of natural waves of a plasma of the 'whistling atmosphere' type, effectively heating electrons and investigated under similar conditions in [276].

5.8. The role of beam electrons in the balance of charged particles of near-electrode plasma

Let us consider the role of high-energy electron beams in the processes of maintaining the balance of the density of charged particles in a near-electrode plasma – the creation and departure of charges.

As in the previous chapter, we will investigate separately the physical conditions in RFCD for the two gas pressure ranges indicated above.

First, we analyze the physical processes for elevated pressures, where no beam-induced plasma instabilities are excited.

The characteristic values of the relaxation lengths of beams with respect to momentum, $l_p \sim 1\text{--}2$ cm, are given above. However, the plasma ionization processes have a larger spatial scale and are characterized by such values as the relaxation time of beam electrons with energy l_e and ionization length λ_{ion}.

We estimate the length l_e by the formula [111]:

$$l_e = \frac{\lambda_e}{\dfrac{2m}{M_i} + \dfrac{v_{ee}}{v_{en}}},$$

where λ_e is the mean free path of plasma electrons; v_{ee}, v_{en} are the frequencies of electron–electron and electron–atom collisions, respectively.

Estimates will be made for two experiments, the results of which are shown in Figs. 4.23, 4.24.

The physical conditions of the first experiment are as follows:

He, $p = 0.5$ Torr, $f = 1.2$ MHz, $V_\sim = 500$ V,

$n_e = 2 \cdot 10^9$ cm^{-3}, $T_e = 5 \cdot 10^4$ K, $l_p = 2.2$ cm.

To calculate the length l_e, we find the values λ_e, v_{ee}, v_{en}.

$$\lambda_e = \frac{1}{pP_c} = \frac{1}{0.5 \cdot 18} = 0.11 \text{ cm}, \quad v_{ee} = \frac{5.5 n_e}{T_e^{3/2}} \ln\left(220 \frac{T_e}{n_e^{1/3}}\right) [277];$$

$$v_{ee} = \frac{5.5 \cdot 2 \cdot 10^9}{\left(5 \cdot 10^4\right)^{3/2}} \ln\left(220 \frac{5 \cdot 10^4}{(2 \cdot 10^9)^{1/3}}\right) = 9.1 \cdot 10^3 \text{ s}^{-1}.$$

$$v_{en} = \frac{v_e}{\lambda_e} = \frac{\left(3kT_e m\right)^{1/2}}{\lambda_e} = \frac{\left(3 \cdot 1.38 \cdot 10^{-16} \cdot 5 \cdot 10^4 \, 9.1 \cdot 10^{-28}\right)^{1/2}}{0.11} = 1.4 \cdot 10^9 \text{ s}^{-1}.$$

$$l_e = \frac{0.11}{\dfrac{2 \cdot 9.1 \cdot 10^{-28}}{6.64 \cdot 10^{-24}} + \dfrac{9.1 \cdot 10^3}{1.4 \cdot 10^9}} \approx 3.9 \cdot 10^2 \text{ cm}.$$

The ionization length of the beam electrons

$$\lambda_{ion} = \frac{1}{pP_{ion}}, \quad P_{ion} = 0.8 \text{ cm}^{-1} \cdot \text{Torr}^{-1} [157]; \quad \lambda_{ion} = \frac{1}{0.5 \cdot 0.8} = 2.5 \text{ cm}.$$

The conditions of the second experiment:

He, $p = 3$ Torr, $f = 1.2$ MHz, $V_\sim = 1$ kV,

$n_e = 3 \cdot 10^{10}$ cm^{-3}, $T_e = 3 \cdot 10^4$ K, $l_p = 0.7$ cm.

Hence we obtain the following values:

$$\lambda_e = 2 \cdot 10^{-2} \text{ cm}, \quad v_{ee} = 2.3 \cdot 10^5 \text{ s}^{-1},$$

$$v_{en} = 6 \cdot 10^9 \text{ s}^{-1}, \quad l_e = 64 \text{ cm}, \quad \lambda_{ion} = 0.8 \text{ cm}.$$

The above estimates show that the length of the electron relaxation in energy is much greater than the length of the discharge gap $d = 6$ cm, i.e. $l_e \gg d$.

Thus, the boundary effects, in particular, the enrichment of the plasma by high-energy beam electrons, cover the entire discharge gap.

As was shown earlier, under the conditions of the experiments, the contribution of the beam electrons to the ionization processes in the near-electrode plasma of the RFCD reached 98%.

Let us now consider qualitatively how the presence of beam 'fast' electrons in the plasma affects the rate at which charges leave the discharge gap. We write down the expressions for the diffusion fluxes of electrons and positive ions, separately writing down the terms for the 'slow' and 'fast' electrons:

$$n_0 v_0 + n_1 v_1 = -D_0 \frac{\partial n_0}{\partial x} - D_1 \frac{\partial n_1}{\partial x} - E b_0 n_0 - E b_1 n_1, \tag{5.1}$$

$$(n_0 + n_1) v_i = -D_i \frac{\partial (n_0 + n_1)}{\partial x} + E b_i (n_0 + n_1), \tag{5.2}$$

where n_0, v_0 are average values of the concentration and directional velocity of the 'slow' electrons; n_1, v_1 – analogous values for the 'fast' electrons; v_i is the directed ion velocity; D_0, D_1, D_i are the diffusion coefficients of 'slow', 'fast' electrons and ions, respectively; b_0, b_1, b_i are the average mobilities of 'slow', 'fast' electrons and ions, respectively.

For the case of ambipolar diffusion in the plasma under consideration, we can write the following expression for the diffusion fluxes of electrons and ions:

$$j_{e,i} = -D_a^B \frac{\partial (n_0 + n_1)}{\partial x}. \tag{5.3}$$

By means of simple calculations analogous to those obtained in the derivation of the expression for the coefficient of ordinary ambipolar diffusion D_a [278], we obtain from equations (5.1), (5.2) the following expression for the 'effective' diffusion coefficient D_a^B, taking into account explicitly the contribution of 'fast' electrons:

$$D_a^B = D_a \left(1 + \frac{n_1 T_1 v_0}{n_0 T_0 v_1} \right), \tag{5.4}$$

Where $\left(D_a = \frac{D_0 b_i + D_i b_0}{b_i + b_0} \right)$, T_0, and T_1 are the 'effective' temperatures of 'slow' and 'fast' electrons, respectively, and v_0 and v_1 are the mean collision frequencies of 'slow' and 'fast' electrons with neutral atoms.

It was assumed that $n_i \approx n_0 + n_1$ and the concentration gradients of 'fast' and 'slow' electrons are quantities of the same order, that is,

$$\frac{1}{n_1} \cdot \frac{\partial n_1}{\partial x} \sim \frac{1}{n_0} \cdot \frac{\partial n_0}{\partial x}.$$

It is clear from (5.4) that

$$D_a^B > D_a.$$

For example, we point out that in one of the cases considered the following orders of magnitude were present:

$$T_0 \approx 5 \, \text{eV}, \quad T_1 \approx 150 \, \text{eV}, \quad n_0 \approx n_1, \quad v_0 \approx v_1.$$

Hence we obtain:

$$D_a^B \approx D_a \frac{150 \, \text{eV}}{5 \, \text{eV}} = 30 D_a.$$

Thus, the presence of the 'fast' beam electrons should lead to a significant intensification of the escape processes of charged plasma particles. The latter facilitates the establishment of the observed 'bell-shaped' distributions of $n_e(x)$ in the vicinity of the electrode (Fig. 3.3).

We also mention here that in the previously proposed discrete mechanism of electron transport through the NESCL to the electrode by means of moving potential wells, the latter should be filled with the 'fastest' electrons of the near-electrode plasma.

In the low-pressure range, due to the appearance of collisionless beam–plasma interaction effects, the physical processes in the near-electrode region acquire a different character. In this case, the relaxation length of the beams by pulse is substantially reduced, and the resulting microwave field of the beam–plasma instability (BPI) further enriches the EES of the plasma with higher energy electrons.

These effects are studied in more detail below, where their strong dependence on the type of gas, anomalously expressed in RFCD in Xe, is also established.

Consider the spatial scales characterizing the ionization processes in low-pressure RFCD.

First we estimate the corresponding lengths in the discharge in He for $p = 2 \cdot 10^{-2}$ Torr, the distributions $I_\lambda(x)$ and $P_\lambda(x)$ in which are shown in Fig. 2.34. The parameters of the near-electrode plasma of a given discharge are $n_e = 5 \cdot 10^9 \, \text{cm}^{-3}$, $T_e = 5 \cdot 10^4$ K. The experimentally determined length $l_p \approx 1$ cm. The values calculated from these data are equal to the mean free path of the plasma electrons $\lambda_e = 3.1$ cm and the beam electrons $\lambda_{eb} = 12.5$ cm, $v_{en} = 5 \cdot 10^7 \, \text{s}^{-1}$, $v_{ee} = 2.2 \cdot 10^4 \, \text{s}^{-1}$, $l_\varepsilon = 4.3 \cdot 10^3$ cm, $\lambda_{ion} = 38.5$ cm.

We make similar estimates for another experiment with RFCD in Ar ($p = 2 \cdot 10^{-2}$ Torr), in which $I_\lambda(x)$ and $P_\lambda(x)$ distributions are obtained (Fig. 2.33). As a result, we obtain: the mean free path of the beam electrons $\lambda_{eb} = 2.8$ cm; the experimentally determined length $l_p = 0.58$ cm; the relaxation length of plasma electrons with respect to energies is $l = 10^3$ cm and the length of the ionization path of the electrons of the beam is $\lambda = 5.9$ cm.

The above estimates show that the ionization processes in the entire discharge gap of the RFCDs under consideration are strongly affected by the boundary effects.

5.9. Beam–plasma instabilities in the RFCD

5.9.1. Experimental data on the presence of beam instability in the examined RFCD

The presence of electron beams in the RFCD experimentally established in the present work raised the question of the effects of the beam–plasma interaction in this discharge.

The author of this book has not come across any papers devoted to the study of beam-plasma instabilities in the RF discharge.

Thus, the study of this problem is very important.

It is well known what wide-ranging practical application was found for RFCD, realized in the form of α- and γ-discharges. There are reasons to believe that the RFCD with the beam–plasma instabilities that have developed in it will allow us to expand the range of parameters of the plasma produced, and, in particular, to acquire new opportunities for the formation of the EES of the gas-discharge plasma.

The experimental results presented here are, in essence, the basis for the work on the creation and study of RF beam-plasma discharge.

The first publications on this subject contain information on the manifestations of BPIs, their mechanisms and parameters, the differences in symmetrical RFCD and asymmetric RFCD, and the significant role of the geometry of the discharge gap [67, 169, 186, 245, 275, 279–289].

The main material for the analysis of the nature of the emerging BPI are the obtained EES of the discharge plasma.

If the role of beam electrons in the ionization of atoms and formation of plasma EES discussed before was directly in the centre of attention, the present discussion will be focused on the participation of microwave fields of beam–plasma instabilities in maintaining the RF discharge.

In connection with what was said above, the discharge under study can be conditionally called RFCD with 'secondary microwave breakdown', implying that three factors are involved in the discharge support: the external RF field, the electron beams, and the secondary microwave field. Considerable attention should be given to the time course of the physical processes in the near-electrode regions of RFCD within the RF field period, caused both by the periodic collapse of the NESCL, and by the time modulation of the parameters of the near-electrode electron beams, which leads to spatial scanning of the discharge gap by the latter.

In the course of the work, a number of facts have been experimentally established, indicating the presence of BPIs effects in the examined RFCD.

First of all, we note the anomalously short relaxation lengths of the NEEB with respect to the momentum, established above with respect to the spatial distribution of the polarization degree of the plasma radiation $P_\lambda(x)$ (Figs. 2.33, 2.34), when the mean free path of the beam in the gas λ_{eb} significantly exceeds the length of the discharge gap d.

It was found that the maximum intensity of plasma radiation is observed near the grounded electrode of RFCD, where the near-electrode electron beam comes from the active electrode and performs significant energy deposition when the $\lambda_{eb} \gg d$ condition is fulfilled (Fig. 5.35).

Under conditions of the presumed presence of BPI, a pulse 1 of chaotic 'hot' electrons, recorded on the oscillogram of electron fluxes to the ground electrode, disappears in the process of NESCL collapse, which destroyed the BPI, after which a pulse 2 of directional beam electrons from the active electrode arises (Fig. 5.30).

Signs of the presence of BPI are also seen in the analysis of resultant plasma EESs. In the case of a symmetrical RFCD, this can be seen from the group of the low-energy electrons, and for asymmetric RFCD – in the high-energy part of the spectrum.

In experiments with a symmetric RFCD in He for $\omega_{0e} \lesssim v_{en}$, where BPI should not develop, it can be seen that with increasing RF voltage the temperature of the slow electrons T_{e1} in the centre of the discharge decreases (Fig. 5.17). This confirms the absence of a significant role of BPI.

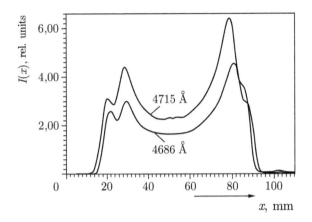

Fig. 5.35. The distribution of the luminescence intensity of the spectral lines along the length of the discharge tube (external electrodes, active electrode on the left). RFCD. f = 2.1 MHz, $V_\sim$ = 500 V.

Under conditions of a symmetric RFCD in He for $\omega_{0e} > v_{en}$ and a monotonic increase of $V_\sim$, the temperature T_{el} initially decreases, and then begins to grow (Fig. 5.20). According to the estimates given below, this corresponds to the expected manifestation of BPI in the investigated discharge regimes.

Predictable with regard to the side effects discussed below, the behaviour of BPI was also observed in the study of EES in a symmetrical RFCD in Xe and air (Figs. 5.21–5.23). For example, according to estimates, in the air under the experimental conditions the threshold of excitation of BPI could not be achieved, which is reflected in the EES obtained, which is very poorly enriched by higher energy electrons (Fig. 5.23).

We note that even relatively weak manifestations of BPI in these experiments were not masked by the effect of stochastic acceleration of electrons at the plasma boundary, since lower field frequencies $f \sim 1$ MHz were used.

As a direct experimental confirmation of the presence of BPI in the investigated discharge, it is possible to consider the behaviour of the beam peak in the plasma EES of asymmetric RFCD, where, due to greatly increased beam densities, the latter was well observed.

As is known from the material presented earlier, the electric field in the NESCL of the RFCD is deeply modulated during the RF field period. Therefore, the parameters of the near-electrode electron beams should also vary considerably with time. In particular, for this reason, one would expect that the beam peak obtained in quasi-stationary measurements in the EES of the discharge plasma will experience an appreciable broadening

The second factor, which greatly broadens the beam peak, should be the beam–plasma instability in the case of its occurrence in the discharge.

Thus, if the first factor were absent or would lead to a slight change in the shape of this peak, then a significant spreading of the latter would clearly indicate the presence of BPI.

Meanwhile, the experiment showed that even a quasi-stationary EES in some discharge modes had a quasi-monoenergetic beam peak (Fig. 5.26).

The latter circumstance can be explained by the fact that the largest contribution to the formation of a signal representing a quasi-stationary EES is made by the phase interval of the RF field period corresponding to the maximum values of the voltage in the NESCL. This has a physical justification that in this interval of phases the ion flux and the radiation intensity of the plasma coming to the electrode and making the density of the emitted electrons, that is, the maximum near-electrode electron beam density, increase. In addition, we note that, given the significant value of the beam peak broadening due to the modulation of the voltage in the NESCL, the half-width of such a peak would significantly exceed all the observed spreading scales of this peak in terms of the energies.

Thus, it has been experimentally established that under the investigated conditions, the voltage modulation in the NESCL does not appreciably broaden the beam peak of the EES observed in the quasi-stationary regime. In the absence of other reasons for the broadening of the beam peak, its energy half-width is $\Delta_{1/2}\varepsilon_e \approx 20–25$ eV (Fig. 5.26), which coincides with the known thermal spread of electrons emitted by the electrode [225].

Thus, the only reason for the observed spreading of the beam peak is the beam plasma instability that arises in the discharge.

Let us pay attention to the two received EES observed in ARFCD in the air at voltages of 600 V and 750 V (Figs. 5.26, 5.27). Here we see an abrupt transition from the EES with a quasi-monoenergetic peak to a spectrum with a spreading peak in the second case, which should be interpreted as the achievement of the threshold for excitation of the beam–plasma instability, based on the estimates given below.

The evolution of the beam peak of the plasma EES of asymmetric RFCD with a monotonically increasing voltage $V_{\sim}$ under conditions more favourable for the excitation of BPI than in the case given above is shown in Fig. 5.28.

These are the qualitative data obtained, which attest to the manifestations of beam–plasma instabilities in the investigated RFCD modes.

5.9.2. Analysis of the characteristics of beam instability

First, let us briefly dwell on some general points of the study of BPI in the examined RFCD.

In connection with various experimental possibilities and specific physical conditions, let us consider separately the characteristics of BPI in the symmetric and asymmetric RFCD.

The object of the investigation was the beam–plasma instability of the Cherenkov type of two species [155]:

1) the dissipative–collisional instability with an increment of growth

$$\delta_1 = \omega_{0e} \left(\frac{n_{eb}}{n_e} \cdot \frac{\omega_{0e}}{v_{en}} \right)^{1/2} \quad \text{and}$$

2) the collisionless instability with an increment

$$\delta_2 = \omega_{0e} \left(\frac{n_{eb}}{2n_e} \right)^{1/3} .$$

The corresponding spatial characteristic, the build-up length of these instabilities

$$L_{1,2} = \frac{u_{eb}}{\delta_{1,2}},$$

where u_{eb} is the velocity of the beam electrons. It is obvious that for the excitation of instabilities the following condition must be satisfied:

$$L_{1,2} < d.$$

The following can be said with respect to the time characteristics of the buildup of BPI.

As is known [155], for effective excitation of BPI under steady-state conditions, the following relation must be satisfied

$$\delta > v_{en}.$$

However, according to what was said above, in the first approximation, the near-electrode electron beams in RFCD exist only during the negative half-period of the RF field $\dfrac{T}{2} \sim \dfrac{\pi}{\omega}$.

Thus, the temporary condition for the development of BPI takes the form:

$$\delta > v_{en}, \quad \frac{\omega}{\pi},$$

that is, the value of δ must exceed the largest of the quantities ω/π and v_{en}. In this case, as can be seen from the expressions for δ_1, δ_2, under the investigated conditions, the ratio $\delta_{1,2} < \omega_{0e}$ was satisfied, since usually [287, 290, 291]: $\dfrac{n_{eb}}{n_e} \sim 10^{-3} - 10^{-2}$ and $\dfrac{\omega_{0e}}{v_{en}} \sim 10^{1} - 10^{2}$.

This means that simply conditions $\omega_{0e} > v_{en}$ are not sufficient for the development of BPI. It is necessary that $\delta > \omega/\pi$, v_{en}.

We note, however, that under the experimental conditions, as a rule, the condition $2\omega \ll v_{en}$ is fulfilled. Hence, in order to excite the beam–plasma instability under our conditions, it is necessary that the following condition be satisfied:

$$\delta > v_{en}.$$

The characteristics of the BPI depend on all external parameters of the RFCD: the kind of gas, pressure, RF voltage, electrode material. The frequency of the RF field is also of considerable importance, since the concentration of the plasma electrons, the drop in the RF voltage in the NESCL and the parameters of the near-electrode electron beam depend on it.

As established experimentally in this work, the geometry of the discharge plays an important role in the excitation of BPI: the diameter of the electrodes, the interelectrode distance and the radius of curvature of the electrodes, which makes it possible to focus the near-electrode electron beams in the discharge gap.

5.9.3. Parameters of the beam–plasma instability in a symmetric RFCD

Since in the investigated symmetric discharge the electrode areas are sufficiently large, the density of the near-electrode beams n_{eb}, respectively, is relatively low. Therefore, it was difficult to study experimentally directly the behaviour of the beam peak of the plasma EES as an indicator of the presence of BPI in various physical conditions.

Thus, it remained to judge the manifestations of instability from the quantitative estimates of the parameters of the BPI and the behaviour of the experimentally determined low-energy part of the EES.

We will carry out an appropriate analysis of the experimental data in various gases.

We have previously presented the obtained electron distribution functions $f_e(\varepsilon)$ (Fig. 5.20) and their representations in the semilogarithmic form $\ln f_e(\varepsilon)$ (Fig. 5.24) for the RFCD in He. It was found that when the voltage was increased in the range $V_\sim = 600-1000$ V, the relative number of fast electrons in the EES and temperature T_{e1} of slow electrons decreased. This was due to an increase in the mean free path of the beams and a purely collisional mechanism for heating the plasma electrons.

The picture changes when $V_\sim \geq 1200$ V, when the temperature T_{e1} rise and the number of electrons of increased energies start to increase.

Such a change in the energy of the electrons upon reaching a certain voltage $V_\sim$, from our point of view, is explained by the appearance of an additional physical process – the excitation of BPI.

In order to justify this, we make a number of estimates of the characteristics of physical conditions.

In the RFCD at He at $p = 0.1$ Torr, the collision frequency of electrons with atoms is $v_{en} = 2.4 \cdot 10^8$ s^{-1}, the minimum observed electron concentrations are $n_{e\,min} \sim 5 \cdot 10^9$ cm^{-3} and, respectively, the plasma frequency $\omega_{0e\,min} = 4 \cdot 10^9$ s^{-1}.

Thus, under all the conditions investigated,

$$\omega_{0e} \gg v_{en},$$

which indicates a low attenuation of plasma waves.

Let us consider the possibility of developing under these conditions beam instabilities of two types: 1) collisional dissipative $(\delta_1 < v_{en})$; 2) collisionless $(\delta_2 > v_{en})$.

We estimate the values of δ_1 and δ_2.

As an example, we carry out calculations for the case: RFCD in He, $p = 0.1$ Torr, $f = 1.2$ MHz, $V_\sim = 600$ V, $d = 6$ cm, $n_e = 5 \cdot 10^9$ cm^{-3}, $\omega_{0e} = 4 \cdot 10^9$ s^{-1}, $T_{e1} = 6.5$ eV, $v_{en} = 2.4 \cdot 10^8$ s^{-1}.

We get:

$$\delta_1 = 4\cdot10^9\left(10^{-3}\cdot\frac{4\cdot10^9}{2.4\cdot10^8}\right)^{1/2} = 5\cdot10^8\ \text{s}^{-1}.$$

Hence it is clear that the condition $\delta_1 < v_{en}$ is not satisfied. Further we have:

$$\delta_2 = 4\cdot10^9\left(\frac{10^{-3}}{2}\right)^{1/3} = 3.2\cdot10^8\ \text{s}^{-1},$$

i.e. $\delta_2 \geq v_{en}$.

Thus, for an amplitude of the RF voltage $V_\sim = 600$ V, the increments δ_1 and δ_2 and the collision frequencies v_{en} are single-order quantities.

Because of this, both named instabilities should arise and the electrons are heated by the fields of waves excited in the plasma.

Apparently, the observed enrichment of the function $f_e(e)$ by 'hot' electrons is due to the combined effect of the ohmic heating of the electrons by the RF field and the collisional dissipative instability (Fig. 5.20).

Table 5.5 shows the dependence of the increments δ_1 and δ_2 on the voltage $V_\sim$ in comparison with the frequency v_{en} for the case under consideration. The obtained experimental dependences of $\frac{\delta_1}{v_{en}}(V_\sim)$ and $\frac{\delta_2}{v_{en}}(V_\sim)$ show that with increasing $V_\sim$ the collisional dissipative instability should weaken and collisionless instability increase accompanied by heating of the beam electrons. The latter effect is found experimentally in this paper and will be discussed below.

Let us compare the development lengths of the beam instabilities $L_{1,2}$ with the distance between the electrodes d.

To estimate the characteristic lengths $L_{1,2}$ we will assume that the beam velocity is equal to $u_{eb} = \left(\frac{2eV_\sim}{m}\right)^{1/2}$, and the increments δ_1 and δ_2 are given in Table 5.5.

Taking into account that the velocity of the beam during the period of the RF field is modulated in magnitude, we note that the lengths $L_{1,2}$ determined in this way correspond to their maximum values. The results obtained show that with increasing $V_\sim$, on the one hand, the heating of the plasma electrons caused by the collisional–dissipative instability must decrease and, on the other hand, conditions for the manifestation of a collisionless instability and an increase in the field strength of plasma waves are improved.

Table 5.5. Dependence of increments δ_1 and δ_2 on RF voltage

V_-, V	v_{en}, 10^{-8} s^{-1}	δ_1, 10^{-8} s^{-1}	δ_2, 10^{-8} s^{-1}	$\dfrac{\delta_1}{v_{en}}$	$\dfrac{\delta_2}{v_{en}}$	L_1, cm	L_2, cm
600	2.4	5.2	3.2	2.1	1.3	4.9	7.9
800	2.4	7.5	4.0	3.1	1.7	3.9	7.4
1000	2.4	9.9	4.9	4.1	2.0	3.3	7.3
1200	2.4	15	6.3	6.2	2.6	2.4	5.7

The observed dependence of the character of the distribution functions $f_e(\varepsilon)$ on the RF voltage (Figs. 5.20 and 5.24) corresponds to the conclusions from the estimates made.

Let us now turn to the experimental results of studying RFCD in Xe (Figs. 5.21 and 5.22). The corresponding characteristics of the physical conditions in the discharge and the values of δ_1, δ_2, and L_1 are presented earlier in Tables 5.2 and 5.3.

In the process of analyzing the data obtained it was assumed that the main mechanism for enriching the plasma EES with 'hot' electrons was the interaction of the near-electrode electron beam with the plasma.

In this case, the Cherenkov collisional–dissipative instability must make the main contribution to the heating of the plasma electrons.

The results of the estimates in Tables 5.2 and 5.3 show that the collisionless instability with an increment δ_2 practically did not develop under these conditions since $\delta_2/v_{en} \leq 1$.

When the RF voltage is monotonously increased, the value of δ_1/v_{en} also increases, which indicates a tendency to transition to a collisionless regime. The latter should lead to a significant decrease in the heating of electrons by the electric field of the instabilities excited by the beams. This is especially evident in Fig. 5.22.

The values of $L_1/d > 1$ given in Tables 5.2 and 5.3 for some RFCD modes do not mean that in principle there is not enough space in the discharge gap for the development of beam instabilities. The point is that the estimates are made for the velocities of the near-electrode electron beams maximum for the period of the RF field. In fact, the velocities of the beams in most of the RF field period are several times smaller than the maximum ones used. Therefore, the values averaged over the period of the RF field $\overline{\left(\dfrac{L_1}{d}\right)} \leq 1$.

Finally, let us consider the results of a study of the EEE of a discharge plasma in air (Fig. 5.23), together with the analysis of the physical conditions for this case given in Table 5.4.

The experimental distributions of $f_e(e)$ exhibit a relatively weak enrichment of high-energy electrons.

As can be seen from Table 5.4, the characteristic values $\delta_1/v_{en} \sim 1$ in the examined RFCD in the air, which means a significant decrease in the intensity of the process of collisional dissipation of the energy of plasma waves excited by electron beams.

In estimating the δ_1/v_{en} ratio, the most probable intervals of the values n_e in the investigated RFCD modes were used on the basis of previous measurements.

Summarizing the results of experimental studies of the EES of the plasma of a symmetric RFCD in He, Xe, and air, we note the following.

The main mechanisms of heating the plasma electrons of these discharges are: 1) heating by the RF field at the active electrode; 2) heating by a field of plasma waves excited by a near-electrode electron beam in the vicinity of a grounded electrode.

Let us consider the trends in the behaviour of these electron heating mechanisms with a monotonic increase in the voltage $V_\sim$ for the gases mentioned above.

He. 1) Heating by an RF field increases with increasing $V_\sim$ and the collision frequency v_{en}. According to the experimental data, $v_{en} \approx$ const and only slightly increases at $V_\sim = 1200$ V. In general, under the investigated conditions, the RF heating was very weakly increased. 2) The heating by the fields of the beam instability is determined by the quantity δ_1/v_{en}, which decreases for $V_\sim = 600–1200$ V, and for $V_\sim \geq 1200$ V is slightly increased. As a result, this kind of heating first decreases noticeably, and at $V_\sim \geq 1200$ V slightly increases. Hence, according to the experimental data, the high-energy part of the distribution $f_e(\varepsilon)$ is determined by the second electron heating mechanism.

Xe. a) $f = 0.6$ MHz.

1) According to the estimated data of Table 5.2, the heating of the electrons by an RF field with increasing $V_\sim$ should increase somewhat, with the exception of case $V_\sim \geq 1200$ V, when the value of v_{en} decreases noticeably.

2) The heating of by the fields of plasma waves decreases, reaching a minimum at $V_\sim = 1200$ V, when the ratio δ_1/v_{en} is maximal, and slightly increases at $V_\sim = 1400$ V.

The non-monotonicity of the dependence of heating of electrons on $V_\sim$ is explained by the manifestation of the Ramsauer effect. The general trend with increasing $V_\sim$ is the depletion of the distribution $f_e(e)$ by the electrons with increased energies.

b) $f = 1$ MHz.

1) As can be seen from the data in Table 5.3, the effective electron temperature $T_{e\ \text{eff}}$ and the collision frequency v_{en} decrease monotonously with increasing $V_\sim$, which indicates a decrease in the heating of electrons by the RF field. In this case, let us pay attention to the fact that at low $V_\sim < 800$ V the main mass of slow electrons is not in equilibrium with the RF field in the plasma, being highly enriched by higher-energy electrons produced by a different heating mechanism.

2) According to the obtained dependence of the distribution function $f_e(\varepsilon)$ (Fig. 5.22) and the ratio δ_1/v_{en} (Table 5.3) on the voltage $V_\sim$, the contribution to the heating of the electrons by the plasma wave fields and the number of fast electrons with increasing $V_\sim$ decrease. Moreover, the depletion of the function $f_e(\varepsilon)$ by the fast electrons occurs in Xe faster than in He. This is explained by the fact that in the first case, in the ratio δ_1/v_{en}, δ_1 is simultaneously increased, and v_{en} decreases, leading to an accelerated transition of the conditions in the discharge from the collision to the collisionless mode.

Air. 1) The specifics of electron heating by an RF field in a given gas consists in depletion of $f_e(\varepsilon)$ by medium-energy electrons ($\varepsilon_e = 6-16$ eV), which, according to the foregoing, is explained by the increased losses of electron energy through additional channels of their interaction with N_2 and O_2 molecules.

2) As noted above, the process of excitation of plasma waves by beams under the conditions studied was weakened due to the fulfillment of the condition $\delta_1/v_{\text{en}} \sim 1$. Therefore, the 'tail' of the distribution $f_e(\varepsilon)$ was relatively poorly enriched in fast electrons.

On the basis of the experimental dependences of $f_e(\varepsilon)$ on $V_\sim$ for all the gases studied, it can be stated that the main factor responsible for the formation of the 'body' of the distribution of $f_e(\varepsilon)$ is the RF field, and the factor determining the formation of the 'tail' is the heating of the electrons by the fields of plasma waves excited by electron beams, that is, as a result, the near-electrode electron beams.

Let us note one more circumstance: the investigated RFCD is an α-type discharge, with a lower RF voltage, i.e., a discharge in which the ionization processes are predominantly determined by the RF field, and not by electron beams.

A characteristic feature of the γ-type RFCD is the very low electron temperature of the plasma ($T_e < 1$ eV).

In our experiments, we usually had $T_e \approx 2-6$ eV. Thus, in contrast to [111], the transition from an α-discharge to a γ-discharge was not achieved. This was due to low frequencies of the RF field and, as a consequence, relatively low values of the conductivity current density, which were less than the normal current density in a DC glow discharge.

Indeed, under the investigated conditions, the length of the ionization range of the near-electrode electron beam substantially exceeded the distance between the electrodes. On the other hand, the condition $\omega_{0e} > v_{en}$ of the weak attenuation of plasma waves has always been fulfilled; the electric field of these waves, in the presence of collisions of electrons with atoms, can significantly heat plasma electrons. Thus, the ionization balance was maintained mainly by thermal electrons.

However, the beams indirectly maintain the ionization balance in the plasma under investigation if the heating of the electrons by the RF field of an external source is commensurable with the heating of the electrons by the electric fields of the emerging instability

5.9.4. The manifestation of BPI in asymmetric RFCD

Due to the fact that in the asymmetric RFCDs the electron beam densities were one or two orders of magnitude larger than in the symmetrical RFCDs, it became possible to record the appearance of the instability directly from the shape of the beam peak of the EES.

Let us consider the quasi-stationary EES of the plasmas of the ARFCD which are shown in Figs. 5.26–5.28.

For a number of reasons, direct measurements of n_e, T_e for the ARFCD were not performed. Therefore, we confine ourselves to physical estimates of plasma parameters and conditions for the development of BPI.

Let us compare the obtained EES of the ARFCD plasma in air at a pressure $p = 0.1$ Torr for $V_\sim = 600$ V (Fig. 5.26) and $V_\sim = 750$ V (Fig. 5.27).

These electron distributions in the high-energy regions are significantly different: at $V_\sim = 600$ V, an almost monoenergetic electron beam is observed, and at $V_\sim = 750$ V its energy range becomes much wider. The spreading of the beam along the energies is due in alkl likelihood to the emerging BPI.

Let us consider the possibility of the development of BPI in ARFCD for the case $V_\sim = 600$ V with the EES, shown in Fig. 5.26.

Let us estimate the increment of the collision–dissipative instability. Under our conditions, $\omega_{0e} \sim 10^9$ s^{-1}; $\frac{n_{eb}}{n_e} \geq 10^{-2}$; $\frac{\omega_{e0}}{v_{en}} \sim 10$ Hence, $\delta_1 \approx 3 \cdot 10^8$ s^{-1}.

Accordingly, for the growth rate of a collisionless instability, we have: $\delta_2 \sim 2 \cdot 10^8$ s^{-1}.

From the data given above we obtain: $\delta_1 \sim \delta_2 \sim v_{en}$.

This means there are unfavourable conditions for the development of both types of BPIs, as a result of which the electron beam should not be appreciably scattered in the fields of these instabilities, and its mean free path in the gas substantially exceeds the interelectrode distance. Thus, the beam should be close to monoenergetic, which was observed experimentally (Fig. 5.26).

The progress of the EES of the plasma to the left of the beam peak is determined by the cascade electrons which arise as a result of collisions of beam electrons with neutral particles, as well as the temporal modulation of the beam parameters.

The EES of the ARFCD plasma at $V_\sim = 750$ V (Fig. 5.27) shows already effective scattering of the electron beam, which is expressed in a significant relative decrease in the amplitude of the beam peak commensurate with the amplitude of the level of the cascade electrons and an increase in its half-width $\Delta_{1/2}\varepsilon_{eb}$ by more than an order of magnitude compared with EES for $V_\sim = 600$ V

$$\frac{\Delta_{1/2}\varepsilon_{eb}(750\ \text{V})}{\Delta_{1/2}\varepsilon_{eb}(600\ \text{V})} = \frac{250\ \text{eV}}{15\ \text{eV}} \approx 17.$$

Obviously, with increasing $V_\sim$, the instability increments δ_1 and δ_2 must increase both in absolute value and in relation to the frequency v_{en}. This means that with increasing $V_\sim$ under these conditions the relation $\delta_2 > v_{en}$ is satisfied. Thus, collisionless instability must be effectively excited, leading to a sharp decrease in the monoenergeticity of the beam [155]. In this case, appreciable heating of the plasma electrons should not occur because of the absence of favorable conditions for the development of the collisional–dissipative instability ($\delta_1 > v_{en}$).

As can be seen from Fig. 5.27, the electron beam at $V_\sim =$ 750 V is not essentially monoenergetic, which can be interpreted as a manifestation of a developed collisionless instability.

Under these conditions, we should expect a significant increase in the density of the near-electrode electron beam, in comparison with the symmetric RFCD, and the transition from the α-discharge to the γ-discharge. Since in the latter the effective temperature of the main electron mass must be very low ($T_{e1} < 0.3$ eV [111]), it is impossible to record a low-energy electron group by an energy analyzer in the investigated ARFCD if there is no appreciable heating of the plasma electrons.

The experimentally studied evolution of the beam peak in the EES of the ARFCD plasma as a function of the RF voltage is shown in Fig. 5.28. This series of measurements was performed for higher voltage values ($850 \leq V_\sim \leq 1000$ V) and lower gas pressure ($p = 0.07$ Torr), compared with the cases considered above.

These conditions are characterized by somewhat larger Q-factors of the resonance effects and an increment of the collisionless instability δ_2.

As can be seen from Fig. 5.28, with increasing $V_\sim$, the beam pulse spreads in energy, and the magnitude of the amplitude ratio of the beam peak to the amplitude of the cascade electron distribution level decreases monotonically to values less than 1.

Thus, under the investigated conditions, the degradation of the near-electrode electron beam increased monotonically with increasing $V_\sim$ under the influence of the collisionless Cherenkov instability, and the character of the high-energy part of the EES was undergoing significant qualitative changes.

5.9.5. RFCD with secondary 'microwave breakdown'

Thus, in the examined RFCD, the appearance of BPI was discovered, which, as is well known, is accompanied by the generation of a microwave field in a plasma.

Earlier it was already reported that electrons with a thermal energy $\varepsilon_e > 620$ eV were found in the impulse flux of electrons to the electrode (Fig. 5.30).

The application of the transverse magnetic field to the discharge gap led to an increase in the flux of high-energy electrons to the electrode monotonically increasing with increasing intensity H_0 (Fig. 5.34).

A similar phenomenon was observed in Ref. [276] when a magnetic field was applied to a plasma region interacting with high-energy stationary electron beams and was explained by the excitation in the plasma of eigenwaves in the microwave range of the 'whistling atmosphere' type effectively heating the plasma electrons.

The flux of 'hot' electrons in RFCD then quickly disappeared, which is explained by the failure of BPI as a result of the specific time course of physical processes in the near-electrode plasma.

The appearance of a significant number of electrons heated by the fields of BPI significantly increases the intensity of ionization processes in the plasma and the conductivity of the discharge gap. The latter can be qualified as a 'secondary microwave breakdown' in the discharge under study.

In the investigated symmetric RFCD, as follows from the data given in the Tables 5.1–5.5, the conditions for excitation of BPI were not optimal. Therefore, the fast-electron enrichment of the discharge function $f_e(\varepsilon)$ in He and in the air was rather weak (Figs. 5.20 and 5.23).

More favourable conditions for the excitation of BPI were observed in the RFCD in Xe. Therefore, the EES of the plasma of this particular discharge was abnormally enriched in the region of increased energies, and the effect of 'secondary microwave breakdown' was observed at moderate RF voltages ($V_\sim = 600$ V). However, with increasing $V_\sim$, due to the Ramsauer effect, the heating of plasma electrons due to the collision–dissipative beam instability rapidly decreased (Fig. 5.22).

In asymmetric RFCD in the air, investigations of the high-energy part of the EES showed significant heating of the beam electrons (an increase in the half-width of the beam peak) with increased RF voltages (Fig. 5.28).

The ability to observe the beam peak in the EES of the ARFCD plasma made it possible to obtain more detailed information on the characteristics of the BPI in the discharge, including data on the temporal course of the development of the instability, its successive time stages – hydrodynamic and kinetic [155].

As can be seen from the oscillograms of the fluxes of charged particles to the grounded electrode of an asymmetric RFCD (Fig. 5.30), the electron flux to the negative half-period of the RF field varies non-monotonically with time. In this case, it has been established by superposition of a transverse magnetic field that pulse

1 is due to 'hot' chaotic electrons, and pulse 2 – by directional beam electrons with energies $\varepsilon_{eb} \sim 2.1$ keV.

The electron pulse 1 sharply decreased in amplitude before the appearance of the pulse 2.

Thus, the process of generation of 'hot' chaotic electrons was observed only in the interval of the phases $\pi - \dfrac{3\pi}{2}$ of the RF field period. Then, the generation of 'hot' electrons was abruptly cut off, and only the high-energy directed electrons remained – the electron beam.

The following explanation of these experimental results is proposed.

In the positive half-cycle of the RF field, the electrons of the near-electrode plasma are practically blocked by a potential barrier in front of the grounded electrode, and there is no electron beam from the active electrode. At the end of the positive half-cycle of the RF field, an electron beam from the active electrode begins to enter the neighbourhood of the grounded electrode whose density and electron energy increase monotonically.

Earlier estimates showed that a BPI can develop in the negative half-period of the field in the near-electrode plasma of a grounded electrode and the increments of this BPI are 2–3 orders of magnitude higher than the frequency ω of the discharge-maintaining field. Because of this BPI should be excited in a very short time $\tau_{1,2} \ll T = 2\pi/\omega$, leading to heating of plasma electrons.

The height of the near-electrode potential barrier decreases simultaneously with the process of electron heating. The latter circumstance leads to an intensification of the process of leaving the electrons to a grounded electrode, amplified by additional heating of the electrons.

The decrease in the density of the near-electrode plasma due to the collapse of the NESCL leads to a decrease in the increments δ_1, δ_2 and to the growth of the instability development lengths L_1, L_2, which may exceed the interelectrode distance d.

The consequence of these processes is the failure of BPI.

With the developed instabilities, the near-electrode electron beam decays rapidly, practically not reaching the grounded electrode. As the excitation process of BPI weakens, its attenuation decreases. As a result, in the second half of the negative half-cycle, an impulse 2 appears on the electrode on the oscillogram of the electron beam as a result of the electron beam reaching the electrode.

As can be seen from the EES of the ARFCD plasma obtained after the breakdown of Cherenkov instabilities, the beam is poorly dispersed in energy (Fig. 5.32), which confirms what was said earlier.

We make the appropriate quantitative estimates.

Let us consider the characteristic high-energy part of the EES of the ARFCD plasma obtained for the phase $\varphi = 3\pi/2$ of the RF field period (Fig. 5.32).

Let us analyze the state of the beam–plasma instability in this phase of the field when a beam pulse 2 is observed on the oscillogram of the electron beam to the electrode (Fig. 5.30).

As is known [155], the hydrodynamic stage of the Cherenkov instability terminates when the spreading range with respect to the velocity of a monoenergetic electron beam v_{Tb} becomes equal to the characteristic value $u\left(\dfrac{n_{eb}}{2n_e}\right)^{1/3}$, where u is the initial velocity of the beam.

For the case under consideration, we estimate the energy of the electron beam ε_{eb}:

$$\varepsilon_{eb} = 2eV_{\sim} = 2 \cdot \sqrt{2} \cdot 10^3 = 2840 \text{ eV}.$$

where $V_{\sim}$ is the effective value of the voltage.

From here:

$$u = \left(\frac{2\varepsilon_{eb}}{m}\right)^{1/2} = \left(\frac{2 \cdot 2840 \cdot 1.6 \cdot 10^{-12}}{9.1 \cdot 10^{-28}}\right)^{1/2} = 3.2 \cdot 10^9 \text{ cm/s.}$$

The energy half-width of the beam peak of the EES is here $\Delta_{1/2}\varepsilon_{eb} \approx 25$ eV. Thus, we get $\Delta_{1/2}\varepsilon_{eb} = \dfrac{mv_{Tb}^2}{2}$ and then:

$$v_{Tb} = \left(\frac{2\Delta_{1/2}\varepsilon_{eb}}{m}\right)^{1/2} = \left(\frac{2 \cdot 25 \cdot 1.6 \cdot 10^{-12}}{9.1 \cdot 10^{-28}}\right)^{1/2} \approx 3 \cdot 10^8 \text{ cm/s.}$$

The characteristic value

$$u\left(\frac{n_{eb}}{2n_e}\right)^{1/3} = 3.2 \cdot 10^9 \left(\frac{1}{2} \cdot 5 \cdot 10^{-3}\right)^{1/3} \approx 4 \cdot 10^8 \text{ cm/s,}$$

where it is accepted that $n_{eb}/n_e \sim 5 \cdot 10^{-3}$, according to the previously mentioned data.

As a result, we obtain $v_{Tb} \approx u \left(\dfrac{n_{eb}}{2n_o} \right)^{1/3}$, which can be interpreted as evidence of the completion of the hydrodynamic stage and the transition to the kinetic stage of the Cherenkov instability at this moment time.

However, another explanation is also possible for the mechanism of appearance of the observed type of the beam peak of EES. The point is that the near-electrode electron beam must not initially be monoenergetic, since it is formed from electrons emitted by the electrode with a certain energy spread. According to [225], the energy spread of these electrons is just $\Delta \varepsilon_e \sim 20-25$ eV. This allows us to assume that in the phase $\varphi = 3\pi/2$ there is no Cherenkov instability (after its collapse by the collapse of NESCL) and the beam is practically not spread out by energy. Apparently, this is also confirmed by the absence of a dependence of the energy half-width of the beam $\Delta_{1/2} \varepsilon_{eb}$ on the amplitude of the RF voltage (Fig. 5.32).

For comparison, we note the degree of beam blurring over the energies in the quasi-stationary EES, which characterizes the conditions with a strongly developed Cherenkov instability at $V_\sim = 950$ V, $\varepsilon_{eb} = 2.7$ keV, and $\Delta_{1/2}\, \varepsilon_{eb} = 750$ eV (Fig. 5.28). In this case, $v_{Tb} = 16 \cdot 10^8$ cm/s, and the value of $u \left(\dfrac{n_{eb}}{2n_e} \right)^{1/3} \approx 2.10^8$ cm/s, i.e.

$$v_{Tb} \gg \left(\frac{n_{eb}}{2n_e} \right)^{1/3}.$$

Thus, a strongly pronounced kinetic stage of the Cherenkov instability is observed here, with a large energy spreading of the beam.

In general, the quasi-stationary EES of the investigated ARFCD reflects the situation, mainly for the kinetic stage of the beam-plasma instability.

In fact, the characteristic time of the quasi-linear relaxation of the hydrodynamic stage of the Cherenkov instability is [155]:

$$t_{HD} \approx \frac{2}{\sqrt{3}} \left(\frac{2n_e}{n_{eb}} \right)^{1/3} \cdot \frac{1}{\omega_{0e}} \approx 10^{-8} \text{ s,}$$

i.e

$$t_{HD} \ll T = \frac{2\pi}{\omega}.$$

Thus, the hydrodynamic stage very quickly becomes kinetic.

The time of quasi-linear relaxation of the kinetic stage of the instability [155]:

$$t_{kin} \approx \frac{1}{\omega_{0e}} \left(\frac{n_e}{n_{eb}} \right)^{4/3} = \left(\frac{n_e}{n_{eb}} \right) \cdot t_{HD},$$

from which follows:

$$t_{kin} \gg t_{HD}.$$

In the RFCD the parameters of the beam and the near-electrode plasma are continuously changing.

Since the value of t_{kin} is 2–3 orders of magnitude greater than t_{HD}, in the presence of conditions for the existence of the Cherenkov instability, in fact, there is always a kinetic instability stage accompanied by heating of the beam and plasma electrons. In this case, there is a considerable scatter of the thermal velocities of the beam:

$$v_{Tb} \gg u \left(\frac{n_{eb}}{2n_e} \right)^{1/3}.$$

We estimate the possibility of the breakdown of the Cherenkov instability due to the collapse process of the NESCL. We shall proceed from the characteristic development lengths of the instability in the hydrodynamic and kinetic stages $\Lambda_{1,2} \sim u_b/\delta_{1,2}$, assuming that the instability is impossible if Λ exceeds the length of the discharge gap d.

An earlier discussion showed that under the conditions studied $\Lambda_{1,2} \sim d = 6$ cm.

Thus, even a small increase in Λ due to a decrease in the density of the plasma electrons n_e due to the collapse of the NESCL can lead to a breakdown of the instability.

On the basis of the previous consideration, we can approximately represent increments of the corresponding stages of instability in the form:

$$\delta_1 = \omega_{0e} \left(\frac{n_{eb}}{n_e} \cdot \frac{\omega_{0e}}{v_{en}} \right)^{1/2} \approx A n_e^{1/4},$$

$$\delta_2 = \omega_{0e} \left(\frac{n_{eb}}{2n_e} \right)^{1/3} \approx B n_e^{1/6},$$

where A and B are constants.

Assume that the NESCL collapse decreases the value of n_e by an order of magnitude. In this case, the value of δ_1 should decrease by approximately 1.8 times, and δ_2 by 1.5 times. Accordingly, the ??? of the collisional dissipative instability L_1 must increase by a factor of 1.8, and that of the collisionless instability of L_2 should increase by a factor of 1.5.

Thus, as a result of the collapse of the NESCL, the relations L_1, $L_2 > d$ can be completely satisfied and the beam instability can be broken.

Moreover, the collisional instability that heats the plasma electrons collapses faster than the collisionless one, since $\delta_1 \sim n_e^{1/4}$, and $\delta_2 \sim n_e^{1/6}$.

We note that even if the concentration n_e decreases only by a factor of two as a result of the collapse of the NESCL, then the length L_1 increases by 20%, and the length L_2 by 15%, which under the conditions given gives, in absolute terms, $\Delta L_1 \approx 1.2$ cm, $\Delta L_2 \approx 0.8$ cm.

The above estimates show that, for admissible changes in n_e due to the collapse of the NESCL, the Cherenkov instability failure is quite possible.

The question of the time course of the EES of the plasma of RFCD during the RF field period is currently poorly known. In the literature there is practically no data on experimental studies of the mechanisms responsible for the temporal course of EES.

Recently, there was only one paper [89] devoted to the experimental study of the time dependence of EES in RFCD using an energy analyzer. The measurements were also carried out in the vicinity of the RF phase of $\varphi = 3\pi/2$, however, at a higher frequency of the RF field (13.56 MHz) and low gas pressures (Ar, $p = 0.004-0.018$ Torr).

An analysis of the mechanisms of the formation of the EES in the above work was not carried out, with the exception of the hypothesis of possible spatial focusing of high-energy electrons in a discharge that is not supported by any quantitative estimates and the necessary data on the parameters of the discharge plasma.

The above proposed physical model of the mechanisms of the formation of EES of the near-electrode plasma of RFCD seems to qualitatively explain the character of the time evolution of EESs obtained in [89].

In conclusion, we note that RFCD with 'secondary microwave breakdown' gives additional control over the parameters of the gas-discharge plasma.

It seems possible to make the transition from the classical RFCD to the RF beam-plasma discharge supported by the electric fields of the beam–plasma instabilities.

It seems possible to make the transition from the classical RFCD to the RF beam–plasma discharge, wholly supported by the electric fields of the beam-plasma instabilities. For example, the experimentally observed spatial structure of the emission analogous to the emission of an asymmetric RFCD (see Fig. 5.36 on the right) is presented in [295], containing periodic quasi-spherically symmetric layers around a brightly luminous nucleus near an active electrode with a small area.

Previously, using the computer simulation of a traditional beam–plasma discharge with an electron gun [296], it was established that periodic ionic layers with variable density should arise from the central region of the discharge in the direction of its periphery.

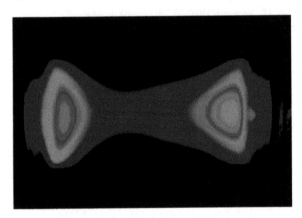

Fig. 5.36. RFCD. Ne, p = 0.5 Torr, f = 6 MHz, $V_\sim$ = 2.1 kV. Active electrode 5 mm in diameter on the right.

References

1. Kaptsov N.A., Electronics. - Moscow: GITTL, 1956.
2. Francis G., Ionization phenomena in gases. - Moscow: Atomizdat, 1964.
3. Chapman B., Glow discharge process - sputtering and plasma etching. - N.Y.: Willey-Interscience, 1980.
4. Raiser Yu.P., Physics of gas discharge. - Dolgoprudny: The Intellekt Publishing House, 2009.
5. Lieberman M.A., Lichtenberg A.J., Principles of Plasma Discharges and Materials Processing. - N.Y.: Willey, 1994.
6. Raiser Yu.P., Shneyder M.N., Yatsenko N.A., High-frequency capacitive discharge. - Moscow: Nauka, 1995.
7. Chabert P., Braithwaite N., Physics of Radio-Frequency Plasmas. - Cambridge: University Press, 2011.
8. Norstrom H., Vacuum. 1980. V. 29, No. 10. P. 341.
9. Hittorf W., Wied. Ann. 1884. V. 52. P. 473.
10. Tesla N., Electr. Engnr. July 1891.
11. Thomson J.J., Phil. Mag. 1891. V. 32. P. 321; V. 32. P. 445.
12. Banerji D., Ganguli, R., Phil. Mag. 1931. V. 11. P. 410.
13. Banerji D., Ganguli, R., Phil. Mag. 1932. V. 13. P. 495.
14. Beck H., Z. Phys. 1935. V. 97. P. 355.
15. Johnson E., Malter L., Phys. Rev. 1950. V. 80. P. 58.
16. Biberman LM, Panin BN, Zh. Teor. Fiz. 1951. Vol. 21. S. 2.
17. Brasefield C.J., Phys. Rev. 1930. V. 35. P. 92.
18. Townsend J.S., Llevellyn Jones F., Phil. Mag. 1931. V. 11. P. 679.
19. Townsend J. S., Llevellyn Jones F., Phil. Mag. 1931. V. 12. P. 815.
20. Brown S.C., Handbuch der Physik / Ed. S. Flugge. - Heidelberg, 1956. - V. 22.
21. Kihara T., Rev. Mod. Phys. 1952. V. 24. P. 45.
22. Llevellyn Jones F., Rep. Progr. Phys. 1953. V. 16. P. 216.
23. Levitsky S.M., Zh. Teor. Fiz. 1957. T. 27. S. 970; C. 1001.
24. Radiophysical Electronics. Ed. N.A. Kaptsov. - Moscow: Moscow State University, 1960.
25. Perel V.I., Pinsky V.I., Zh. Teor. Fiz. 1963. V. 33. P. 33.
26. Chenot M., C. Roy. Acad. Sci. 1952. V. 234. P. 608.
27. Dzherpetov H.A., Pateyuk G., JETP. 1955. V. 28. P. 3.
28. Butler H., Kino G., Phys. Fluids. 1963. V. 6. P. 1346.
29. Dzharpetov Kh.A., Bulkin P.S., Akhmedov A.R., Vestn. Moscow. un-ta. Ser. Fizika i astronomiya. 1959. No. 3. P. 38.
30. Allis W.P., Brown S.C., Everhart E., Phys. Rev. 1951. V. 84. P. 519.
31. Kojima S., Takayama K., Shimachi S., J. Phys. Soc. Japan. 1953. V. 8. P. 55.
32. Solntsev G.S., Porokhin A.G., Chistyakova N.M., Izv. AN SSSR. 1959. V. 23. P. 46.
33. Bayet M., Guerineau F., C. R. Acad. Sci. 1954. V. 239. P. 1029.

34. Fransis G., Pros. Phys. Soc. 1955. V. B68. P. 137.
35. Hatch A.J., Williams H.B., Phys. Rev. 1953. V. 89. P. 339; 1955. V. 100. P. 1228.
36. Kucherenko V.I., Fedorus A.I., Radiotekhnika i elektronika. 1959. V. 4. P. 89.
37. Pateyuk G.M., Dissertation. Cand. fiz.-mat. sciences. - Moscow: Moscow University.1953.
38. Flugge W., Ann. der Phys. 1956. V. 18. P. 251.
39. Schneider F., Z. Ang. Phys. 1952. V. 4. P. 324; 1954. V. 6. P. 456.
40. Gagne R.R., Cantin A. J. Appl. Phys. 1972. V. 43. P. 2639.
41. Vagner S.D., Virolainen J.F., Zh. Teor. Fiz. 1968. P. 38. S. 45; Opt. i spektr. 1970. 28. 28. P. 32.
42. Krindach D.P., Tunitsky N.F., Zh. Teor. Fiz. 1968. P. 38. P. 79.
43. Motornenko P., Truten N., Opt. i spektr. 1965. P. 18. P. 1074.
44. Bochkova O.P., Razumovskaya L.P., Opt. i spektr. 1965. P. 18. P. 777.
45. Godyak V.A., Kuzovnikov A.A., Savinov V.P., El Sammani A.J., Vestn. Moscow. un-ta. Ser. Fizika i astronomiya. 1968. No. 2. 126.
46. El Sammani A.J. Dissertation. Cand. fiz.-mat. sciences. - Moscow: Moscow University, 1969.
47. Kuzovnikov A.A., Savinov V.P., Proc. II All-Union. Conf. of Physicist of Low Temp. Plasma. - Minsk, 1968. - P. 190.
48. Kuzovnikov A.A., Savinov V.P., Problems of low-temperature plasma physics. - Minsk, 1970.
49. Savinov V.P., Dissertation. Cand. fiz.-mat. sciences. - Moscow: Moscow University, 1970.
50. Kuzovnikov A.A., Savinov V.P., Radiotekhnika i elektronika. 1973. V. 18. P. 816.
51. Kuzovnikov A.A., Savinov V.P., Vestn. Moscow. un-ta. Ser. Fizika i astronomiya. 1973. No. 2. P. 215.
52. Boedeker H., Haldemann G., AIAA J. 1968. V. 6. P. 92.
53. Koenig H. R., Maissel L.I., IBM J. Res. Devel. March 1970. V. 14. P. 168.
54. Kuzovnikov A.A., Godyak V.A., Fizika plazmy. 1975. Vol. 1. P. 496.
55. Godyak V.A., Kuzovnikov A.A., Khadir M.A., Proc. II All-Union. Conf. of Physicist of Low Temp. Plasma. - Minsk, 1968. - P. 92.
56. Godyak V.A., Zh. Teor. Fiz. 1971. P. 41. P. 1364.
57. Andreev A.D., Vestn. Belorussk. un-ta. Ser. 1. 1969. No. 2. 78.
58. Haydon S.C., Plumb I.C., J. J. Phys. D: Appl. Phys. 1978. V. 11. P. 1721.
59. Norstrom H., Vacuum. 1980. V. 29. P. 341.
60. Norstrom H., Blom H.O., Thin Sol. Films. 1981. V. 86. P. 255.
61. Tesar C., Acta phys. slov. 1979. V. 29. P. 109.
62. Popov O.A., Godyak V.A., J. Appl. Phys. 1985. V. 57. P. 53.
63. Bletzinger P., Flemming, M.J., J. Appl. Phys. 1987. V. 62. P. 4688.
64. Goduak V.A., Piejak R. B., Alexandrovich B.M., Rev. Sci. Instr. 1990. V. 61. P. 2401.
65. Bletzinger, P., De Joseph C.A., IEEE Trans. Plasma Sci. 1986. V. PS-14. P. 124.
66. Yatsenko N.A., Zh. Teor. Fiz. 1988. Vol. 58. P. 294.
67. Kovalevsky V.L., Savinov V.P., Fizika plazmy. 1994. Vol. 20. P. 322.
68. Raizer Yu.P., Schneider M.N., J. Phys. D: Appl. Phys. 1994. V. 27. P. 1457.
69. Kovalevsky V.L., Savinov V.P., Singaevsky I.F., Proc. Conf. Physics of low-temperature plasma. - Petrozavodsk, 1995. Part 2. P. 243.
70. Kuzovnikov A.A., Kovalevsky V.L., Savinov V. P., Yakunin V. G., J. de Phys. 1979. V. 40. Suppl., Coll. C7. P. 459.
71. Raiser Yu.P., Schneider M.N., Fizika plazmy. 1987. V. 13. P. 471.

72. Chung C., Zhong J., Whittaker C. J. Appl. Phys. 1987. V. 62. P. 1633.
73. Wood B. P., Lieberman M. A., Lichtenberg A. J., IEEE Trans. Plasma Sci. 1991. V. 19. P. 619.
74. Yatsenko N.A., TVT. 1982. Vol. 20. P. 1044.
75. Kovalevsky V.L., Kondrakov R.E., Savinov V.P., Singaevsky I.F., Proc. Conf. Physics and technology of plasma." - Minsk, 1994. - V. 1. P. 373.
76. Gurin A.A., Chernova N.N., Fizika plazmy. 1985. P. 11. P. 244.
77. Derouard J., Boeuf J.P., Proc. of ESCAMPIG, 1990. P. 12.
78. Raizer Yu.P., Shneider M.N., Fizika plazmy. 1992. P. 18. P. 1211.
79. Kuzovnikov A.A., Kovalevsky V.L., Savinov V.P., Proc. VIII All-Union. Conf. Generation of low temperature plasma. - Novosibirsk, 1980. - Part 3. P. 370.
80. Pointy A., Appl. Phys. Lett. 1987. V. 50. P. 316.
81. Savinov V.P., Yakunin V.G., Proc. ICPIG-XIX. Belgrade, 1989. V. 1. P. 123.
82. Gordeev A.I., Kovalevsky V.L., Savinov V.P., Shorin A.B., Proc. ICPIGXX. Piza, 1991. P.V. P. 83.
83. Meijer, P.M., Goedheer W. J., Proc. ESCAMPIG. 1990. P. 129.
84. Goedheer W.J., Manenshijn A., Proc. ESCAMPIG, 1990. P. 80.
85. Hebner, G., Kushner, M., J. Appl. Phys. 1987. V. 62. P. 2256.
86. Kohler K., Horne D.E., Coburn J.W., J. Appl. Phys. 1985. V. 58. P. 3350.
87. Bohm C., Perrin J., Rev. Sci. Instr. 1993. V. 64. P. 31.
88. Smirnov A.S., Ustavshchikov A.Yu., Frolov K.S., Proc. Conf. Physics and technology of plasma. - Minsk, 1994. - v. 2. - P. 120.
89. Smirnov A.S., Frolov K.S., Usavtchikov A.Yu., Proc. conf. Low-temperature physics. plasma. - Petrozavodsk, 1995. - V. 2. - P. 178.
90. Drachev A.I., Kazantsev S.A., Rys A.G., Subbotenko A.V, Optika i spektr. 1988. V. 64. P. 1185.
91. Wendt H., Hitchon S., J. Appl. Phys. 1992. V. 71. P. 4718.
92. Hasper W., Bohm H., Horchauer B., J. Appl. Phys. 1992. V. 71. P. 4168.
93. Godyak V.A., Popov O.A., Ganna A.S., Radiotekhnika i elektronia. 1976. V. 21. P. 2639.
94. Kovalevsky V.L., Savinov V.P., Urieva I.A., Proc. II All-Union. Conf. on the Physics of gas discharge. - Makhachkala, 1988. - Part 1. - P. 98.
95. Wagner S.D., Kagan Yu.M., Niskonen I.S., Zh. Teor. Fiz. 1977. V. 47. P. 772.
96. Volkova E.A., Popov A.M., Popovicheva O.V., Rakhimova T.V., Fizika plazmy. 1990. Vol. 16. P. 738.
97. Gottsho R.A., Donnelly V.M., J. Appl. Phys. 1984. V. 56. P. 245.
98. Dyatko N.A., Kochetov N.V., Napartovich A.P., Fizika plazmy. 1985. V. 11. P. 739.
99. Makabe T., Nakaya M., J. Phys. D: Appl. Phys. 1987. V. 20. P. 1243.
100. Kapicka V., Folia Facult. Sci. Natur. Univ. 1976. Physika 22. V. 17. P. 5.
101. Lyutyi A.I., Melnikova L.D., Sokolova N.L., Izv. VUZ. Fizika. 1975. No. 1. P. 56.
102. Cox T.J., Deshmukh V.G., Hope D. A., et al. J. Phys. D: Appl. Phys. 1987. V. 20. P. 820.
103. Ostapchenko E.P., Oskin V.A., Ryumin V.B., Stepanov V.A., Kvant. elektronika. 1976. V. 3. S. 1980.
104. Vasquie S., Bacri J., Benot-Cattin P., Blanc D., Proc. ESCAMPIG. 1990. P. 226.
105. Godyak V.A., Oks S.N., Zh. Teor. Fiz. 1979. No.7.
106. Godyak VA, Oks S.N., Zh. Teor. Fiz. 1979. No.10.
107. Belenguer Ph., Boeuf J.P., Proc. ESCAMPIG. 1990. P. 226.
108. Vasquie S., Bacri J., Blanc D., Clavery F., Nuovo Cimento. 1969. V. 62B. No. 1.
109. Shinna Y., Yoshihiro O., Yassunori O., Hiroharu F., Japan J. Appl. Phys. 1992.

110. Stadler K., Anderson C., Multan A., Graham W., J. Appl. Phys. 1992. V. 72. P. 245.
111. Goduak V. A., Piejak R. B., Alexandrovich B.M., Plasma Source Sci. Technol. 1992. V. 1. P. 36.
112. Kortshagen U., J. Phys. D: Appl. Phys. 1993. V. 26. P. 1230.
113. Kimura T., Oke K., Japan J. Appl. Phys. 1993. V. 32. P. 3601.
114. Curran, J. E., Vacuum. 1984. V. 34. P. 343.
115. Dilonardo M., Capitelli M., Winkler R., Wilgelm J., Mat. Res. Soc. Symp. Proc. 1986. V. 68. P. 287.
116. Goduak V. A., Piejak R. B., Alexandrovich B.M., Phys. Rev. Lett. 1992.V.68. P. 40.
117. Kuzovnikov A.A., Savinov V.P., Yakunin V.G., Proc. II All-Union meeting on the Physics of electrical breakdown of gases. - Tartu, 1984. - Part 1. - P. 114.
118. Khaselev V.Ya., Mikhalevsky V.S., Tolmachev G.N., Fizika plazmy. 1980. V. 6. S. 430.
119. Dilecce G., Capitelli M., De Benedictis S. J. Appl. Phys. 1991. V. 69, No. 1. P. 4168.
120. Kline L.E., Kushner M.J., Solid State and Mater. Sci. 1989. V. 16. P. 1.
121. Smirnov A.S., Tsendin L.D., Proc. of ICPIG-XIX. Belgrade, 1989. V. 2. P. 456.
122. Smirnov A.S., Tsendin L.D., IEEE Trans. Plasma Sci. 1991. V. 19. P. 130.
123. Kuzovnikov A.A., Kovalevsky V.L., Savinov V.P., Kurzfassungen der Vortrage V Tagung "Physik und Technik des plasmas". - Greifswald, 1980. - P. 107.
124. Barnes M. S., Colter F. J., Elta M. E., J. Appl. Phys. 1987. V. 61. P. 81.
125. Kaganovich I.D., Tsendin L.D., IEEE Trans. Plasma Sci. 1992. V. 20. P. 66; P. 86.
126. Rogoff G.L., Kramer J.M., Piejak R. B., IEEE Trans. Plasma Sci. 1986. V. PS-14. P. 103.
127. Pointy A.M., J. Appl. Phys. 1986. V. 60. P. 4113.
128. Goduak V.A., Khanneh A.S., IEEE Trans. Plasma Sci. 1986. V. PS-14. P. 112.
129. Latush E.L., Mikhalevsky V.S., Sam M.F., and others., Pis'ma ZhETF. 1976. Vol. 24. P. 81.
130. Norstrom H., Vacuum. 1980. V. 29. P. 443.
131. Kovalev A.S., Nazarov A.I., Rakhimov A.T. and others, Fizika plazmy. 1986. V.12. P. 1264.
132. Yatsenko N.A., Izv. RAS. Ser. fiz. 1992. P. 56. P. 77.
133. Lipatov N.I., Pashinin P.P., Prokhorov A.M., Yurov V.Yu., Proceedings IOFAN. 1989. Vol. 17. S. 53.
134. Yatsenko N.A., Pis ma Zh. Teor. Fiz. 1993. T. 19. P. 72.
135. Steinbruchel C., Curtis B.J., Lehmann H.W., Widmer R., IEEE Trans. Plasma Sci. 1986. V. PS-14. P. 137.
136. Overzet L.J., Beberman J. H., Verdeyen J. T., J. Appl. Phys. 1989. V. 66. P. 1622.
137. Van Roosmalen A., Van den Hoeg W., Kalter H., J. Appl. Phys. 1985. V. 58. P. 653.
138. Gill M., Vacuum. 1984. V. 34. P. 357.
139. Kuzovnikov A.A., Savinov V.P., Yakunin V.G., Vestn. Moscow. un-ta. Ser. Fizika i astronomiya. 1980. V.21. P. 75.
140. Kuzovnikov A.A., Savinov V.P., Novoselov A.N., Yakunin V.G., Radiotekhnika i elektronika. 1980. V. 25. P. 1677.
141. He D., Hall D.R., IEEE J. Quant. Electr. 1984. V.QE-20. P. 509.
142. Flamm D.L., J. Vac. Sci. Technol. 1986. V.A4. P. 729.
143. Sinclair L.R., Tulip J., J. Appl. Phys. 1984. V. 56. P. 2497.
144. Ilic D.B., Rev. Sci. Instr. 1981. V. 52. P. 1542.
145. Savinov V.P., Dissertation. Doctor. fiz.-mat. sciences. - Moscow: Moscow State University, 2001.
146. Lieberman M.A., Booth J. P., Chabert, P., et al., Plasma Source Sci. Technol. 2002.

V. 11. P. 283.

147. Aleksandrov A.F., Rukhadze A.A., Prikl. fizika. 1995. No. 2. P. 56.
148. Lisovskiy V.A., Booth J.-P., Landry K., et al., Europhys. Lett. 2008. V. 82, No. 1. P. 15001.
149. Lisovskiy V.A., Booth J.-P., Landry K. et al. J. Phys. D: Appl. Phys. 2006. V. 39, No. 4. P. 660.
150. Kitajima T., Takeo Y., Petrovich Z., Makabe T., Appl. Phys. Lett. 2000. V. 77. P. 489.
151. Granovsky V.L., Electric current in the gas. - Moscow: GITTL, 1952.
152. Babat G.I., Vestn. Elektropromyshl.. 1942. No.2. P. 1.
153. Physics and Technology of Low-Temperature Plasma, Ed. S.V. Dresvin. - Moscow: Atomizdat, 1972.
154. Mac-Kinnton, K.A., Phil. Mag. 1929. V. 8. P. 605.
155. Aleksandrov A.F., Bogdankevich L.S., Rukhadze A.A., Fundamentals of plasma electrodynamics. - Moscow: Vysshaya shkola, 1988.
156. Velikhov E.P., Kovalev A.S., Rakhimov A.T., Physical phenomena in the gas-discharge plasma. - Moscow: Nauka, 1987.
157. Brown S., Elementary processes in the plasma of a gas discharge. - M .: Gosatomizdat, 1961.
158. Khovatson A.M., Introduction to the theory of gas discharge. - Moscow: Atomizdat, 1980.
159. Savinov V.P., The study of gas discharge: Textbook. Moscow State University, 2006.
160. Yatsenko N.A., Zh. Teor. Fiz. 1980. V. 50, No. 11. P. 2480.
161. Lisovsky V.A., Zh. Teor. Fiz. 1998. V. 68, No. 5. P. 52.
162. Raizer Yu.P., Shneyder M.N., Fizika plazmy. 1987. Vol. 13, No. 4. P. 471.
163. Bohm C., Perrin J., J. Phys. D. 1991. V. 24, No. 6. P. 865.
164. Kropotov N.Yu., Lisovsky VA, Kachanov Yu.A., and others., Pis'ma Zh. Teor. Fiz. 1989. Vol. 15, No. 21. P. 17.
165. Engel A., Shteenbeck M., Physics and Technology of Electric Discharge in gases / Ed. N.A. Kaptsov. - M.-L .: ONTI, 1936.
166. Raizer Yu.P., Surzhikov S.T., Teplofiz. Vys. Temp.. 1988. V. 26. P. 428.
167. Raizer Yu.P., Physics of gas discharge. - Moscow: Nauka, 1992.
168. Boeuf J.P., Belenguer P., Non-balanced processes partially ionized gases. Eds. M. Capitelli, N.J.B ardsley. - N.Y .: Plenum, 1990. - P. 155.
169. Aleksandrov A.F., Rukhadze A.A., Savinov V.P., Singaevsky I.F., Pis'ma Zh. Teor. Fiz. 1999. Vol. 25, No. 19. P. 32.
170. Isaev N.V., Rukhadze A.A., Shustin E.G., Fizika plazmy. 2005. V. 31, No.11. P. 1026.
171. Klykov I.L., Tarakanov V.P., Shustin E.G., Fizika plazmy. 2005. V. 38, No.3. C. 290.
172. Lisovsky V.A., FIP. 2006. V. 4, No. 3, 4. P. 143.
173. Chabert P., Raimbault J.-L., Levif P., et al., Plasma Source Sci. Technol. 2006. V. 15. P. 130.
174. Godyak V.A., Soviet Radio Frequency Discharge Research. Delphic Associates Inc. Falls Church VA. 1986.
175. Lieberman M.A., J. Appl. Phys. 1989. V. 65. P. 4186.
176. Lieberman, M.A., Savas S.E., J. Vac. Sci. Technol. 1990. V.A8. P. 1632.
177. Yeom G.Y., Thornton, J.A., Kushner M. J., J. Appl. Phys. 1989. V. 65. P. 3825.
178. Lisovskiy V.A., Yegorenkov V.D. J. Phys. D: Appl. Phys. 1998. V. 31. P. 3349.
179. Levitsky S.M., Zh. Teor. Fiz. 1957. V. 2. P. 887.
180. Lisovskiy V.A., Yegorenkov V.D., J. Phys. D: Appl. Phys. 1994. V. 27. P. 2340.
181. Lisovsky VA, Booth J.-P., Landry K., etc., FIP. 2005. V. 3, No.1-2. P.70.

182. Engel A. Ionized gases. - Moscow: Fizmatgiz, 1959.
183. Chen Sin-Li, Sekiguchi T., J. Appl. Phys. 1965. V. 36. P. 2363.
184. Gottscho R.A., Mandich M.L., J. Vac. Sci. Technol. 1985. V.A3. P. 617.
185. Gottscho R.A., Phys. Rev. A. 1987. V. 36. P. 2233.
186. Alexandrov A.F., Savinov V.P., Singaevsky I. F., Beam-plasma instability effects supporting capacitive low pressure RF discharges, Advanced Technologies Based on Wave and Beam Generated Plasmas. NATO ASI Series. Ser. 3. V. 67. - Kluwer Acad. Publ., 1999. - P. 557.
187. Annaratone B.M., Ku V.P., Allen J.E., J. Appl. Phys. 1995. V. 77. P. 5455.
188. Orlov K.E., Smirnov A.S., Plasma Source Sci. Technol. 2001. V. 10. P. 541.
189. Alexandrov A.F., Kovalevsky V.L., Savinov V.P., Singaevsky I.F., Proc. of ICPIG-22. V. 2. Hoboken, USA, 1995. P. 167.
190. Yatsenko N.A. Dissertation. Doct. fiz.-mat. sciences. - Moscow: IOFAN, 1991.
191. Lisovskiy V.A., Martins S., Landry K., et al., Phys. of Plasmas. 2005. V. 12, No. 9. P. 093505.
192. Lipatov N.I., Pashinin P.P., Prokhorov A.M., Yurov V.Yu., Trudy IOFAN, 1989.
193. Zhukov A.A., Kovalevsky V.L., Kruglov M.S., et al., Uchebnyi eksperiment v obraovanii. 2012. No. 2. P. 29.
194. Kakuta S., Makabe T., Tochikubo F., J. Appl. Phys. 1993. V. 74. P. 4907.
195. Kuzovnikov A.A., Kovalevsky V.L., Savinov V.P., Vestn. Mosk. un-ta. Ser. Fizika, astronomiya. 1983. V. 24. P. 28.
196. Garscadden A., Emeleus K.G., Proc. Phys. Soc. 1962. V. 79. P. 535.
197. Godyak V.A., Kuzovnikov A.A., Proc. Conf. in chemistry and low-temperature physics. plasma. M., 1967. P. 59.
198. Godyak V.A., Fizika plazmy. 1976. v. 2. P. 141.
199. Yatsenko N.A., Proc. All-Union. Meeting High-frequency discharge in wave fields. Gorky, 1987. P. 22.
200. Yatsenko N.A., Proceedings of the Moscow Institute of Physics and Technology. Ser. Obshch. i molekulyar. fizika. Dolgoprudny: MIPT, 1978. P. 226.
201. Kuzovnikov A.A., Kovalevsky V.L., Savinov V.P., Yakunin V.G., Proc. of ICPIG-XIII. Berlin, 1977. P. 1. P. 343.
202. Chistyakov P.N., Physical Electronics: Issue. 3. Ed. P.N. Chistyakov. - MIFI; Atom-izdat, 1966. - P. 3.
203. Kuzovnikov A.A., Dissertation. Doct. fiz.-mat. sciences. - Moscow: Moscow State University, 1969.
204. Aleksandrov A.F., Zh. Teor. Fiz. 1965. V. 35. P. 35; In the same place. P. 226.
205. Tarasova V.V., Dissertation. Cand. fiz.-mat.nauk. - Moscow: Moscow State University, 1969.
206. Aleksandrov A.F., Ryabyy V.A., Savinov V.P., Yakunin V.G., Fizika plazmy. 2002. Vol. 28, No. 12. P. 1086.
207. Bruce R.H. J. Appl. Phys. 1981. V. 52. P. 7064.
208. Kohler, K., Coburn, J.W., Horne, D. E., et al. J. Appl. Phys. 1985. V. 57. P. 59.
209. Metze A., Ernie D.W., Oskam H. J. J. Appl. Phys. 1986. V. 60. P. 3081.
210. Tsui R.T., Phys. Rev. 1968. V. 168. P. 107.
211. Brusilovsky B.A., Kinetic ion-electron emission. - Moscow: Energoatomizdat, 1990. - P. 184.
212. Smirnov A.S., Orlov K.E., Pis'ma Zh. Teor. Fiz. 1997. V.23. P. 39.
213. Smirnov A.S., Tsendin L.D., Zh. Teor. Fiz. 1991. 61. P. 64.
214. Kovalevsky V.L., Kondrakov R.E., Savinov V.P., Singaevsky I.F., Proc.. VIII conf. on the Physics of gas discharges. Ryazan, 1986. Part 2. P. 60.

215. Kagan Yu.M., Perel' V.I., Usp. Fiz. Nauk. 1963. V. 81. P. 409.
216. Kazantsev S.A., Subbotenko A.V., Fizika plazmy. 1984. V. 10. P. 135.
217. Kazantsev S.A., Usp. Fiz. Nauk., 1983. V. 139. P. 621.
218. Vidaud P., Durrani S.M., Hall D.R. J. Phys. D: Appl. Phys. 1988. V. 21. P. 57.
219. Hebner, G. A., Verdeyen, J. T., IEEE Trans. Plasma Sci. 1986. V. PS-14. P. 132.
220. Kovalevsky V.L., Savinov V.P., Singaevsky I.F., Yakunin V.G., Proc. XXIX Scientific Conf. in Physical and mathematical and natural science. Rostov University of the Friendship between nations. - M., 1993. - Part 1. - P. 16.
221. Kovalevsky V.L., Riaby V.A., Savinov V.P., Yakunin V.G., Proc. of ICPIG-XXX. Belfast, Northern Ireland, UK, 2011.
222. Karimov R.G., Murav'ev I.I., Izv. VUZ. Fizika. 1975. No. 6. P. 87.
223. Zapesochnyi I.P., Fel'tsan P.V., Ukr. Fiz. Zh. 1965. V. 10. P. 1197.
224. Bogdanova N.A., Jurgenson S..V., Optija i spektrosk.1986. V. 61. P. 241.
225. Hagstrum H.D., Phys. Rev. 1956. V. 104. P. 672.
226. Malyshev V.I. Introduction to experimental spectroscopy. - Moscow: Nauka, 1979. P. 479.
227. Radiative properties of solid materials. Ed. A.Sh. Sheidlin. - Moscow: Energiya, 1974. P. 215.
228. Golubovsky Yu.B.. Vestn. LGU. 1967. No.10. P. 64.
229. Borodin V.S., Golubovsky Yu.B., Vestn. LGU. 1970. No. 16. P.49.
230. Bogorodsky M.M., Rimashevskaya L.R., Semihokhin E.A., Vestn. MGU. Ser. khim. 1976. V. 17. P. 402.
231. Osherovich A.L., Virolainen Ya.F., Vestn. LGU. 1966. No. 4. 140.
232. Ingmar S.G., Braithwaite N.S., J. Phys. D: Appl. Phys. 1988. V. 21. P. 1496.
233. Appleton E., Childs F., Phil. Mag. 1930. V. 10. P. 969.
234. Aleksandrov A.F., Kuzovnikov A.A., "Izv. VUZ. Radiofizika. 1968. V. 11. P. 1548.
235. Savinov V.P., Yakunin V.G., Proc. of ICPIG-XIX. Belgrad, 1989. P. 220.
236. Shapiro G.I., Soroka A.M., Pis'ma Zh. Teor. Fiz. 1979. V. 5. P. 129.
237. Savinov V.P., Singaevsky I.F., Proc. X Conf. on the Physics of gas discharge. Ryazan, 2000. Part 1. P. 11.
238. Kawamura E., Vahedi V., Lieberman M.A., Birdsall V., Plasma Source Sci. Technol. 1999. V. 8. P. 45.
239. Allis W. P., Rose D. J., Phys. Rev. 1954. V. 93. P. 84.
240. Riemann K.-U. J. Appl. Phys. 1989. V. 65. P. 999.
241. Goedheer W. J., Meijer P.M., Abstracts of ESCAMPIG-X. Orleans, 1990. P. 388.
242. Kazantsev SA, Subbotenko A.V., Pis'ma Zh. Teor. Fiz. 1984. V. 10. P. 1251.
243. Chernykh K.A. Diploma thesis. - Moscow: Moscow State. University, 1991.
244. McDaniel, I., Collision Processes in Ionized Gases. - Moscow: Mir, 1967. - P. 832.
245. Kazantsev S.A., Kovalevsky V.L., Kuzovnikov A.A., etc., Vestn. LGU. 1990. Ser. 4, No.4. P. 26.
246. Ionov N.I., Zh. Teor. Fiz. 1964. V. 34. P. 194.
247. Batanov G.M., Petrov N.N., Fiz. Tverd. Tela. 1959. V. 1. P. 1586.
248. Kazantsev S.A., Svelokuzov A.E., Subbotenko A.V., Zh. Teor. Fiz. 1986. V. 56. P. 1091.
249. Lieberman M.A., IEEE Trans. Plasma Sci. 1988. V. 16. P. 638.
250. Godyak V.A., Piejak R. B., Sternberg N., IEEE Trans. Plasma Sci. 1993. V. 21. P. 378.
251. Sommerer T. J., Hitchon W. N., Harvey R. E., Lowler J. E., Phys. Rev. 1991. V. 43. P. 4452.
252. Brunet A., Taillet J., J. Physique. 1965. V. 26. P. 517.

253. Popov O. A., Godyak V.A. J. Appl. Phys. 1985. V. 57. P. 53.
254. Cooperberg D. J., Birdsall C. K., Plasma Source Sci. Technol. 1998. V. 7. P. 96.
255. Ku V.P.T., Annaratone B.M., Allen J.E., J. Appl. Phys. 1998. V. 85. P. 6536.
256. Holstein T., Phys. Rev. 1946. V. 70. P. 367.
257. Winkler R., Dilonardo M., Capitelli M., Wilhelm J., Plasma Chem. Plasma Proc. 1987. V. 7. P. 125.
258. Makabe T., Goto N., J. Phys. D: Appl. Phys. 1988. V. 21. P. 887.
259. Kushner, M. J., IEEE Trans. Plasma Sci. 1986. V. PS-14. P. 188.
260. Kline, L. E., Partlow, W. D., Bies, W. E., J. Appl. Phys. 1989. V. 65. P. 70.
261. Gorbunov N.A., Iminov K.O., Kudryavtsev A.A., Zh. Teor Fiz. 1988. Vol 58. P. 2301.
262. Capitelli M., Celiberto R., Course C. et al. J. Appl. Phys. 1987. V. 62. P. 4398.
263. Shi B., Fetzer G.I., Yu. Z., et al., IEEE J. Quant. Electr. 1989. V. 25. P. 948.
264. Gill P., Webb C.E.E., J. Phys. D: Appl. Phys. 1977. V. 10. P. 299.
265. Winkler R., Capitelli M., Gourse C., Wilhelm J., Proc. of ESCAMPIGVIII. Orleans, 1990. P. 390.
266. Surendra M., Graves D.B., Morey I.J., J. Appl. Phys. Lett. 1990. V. 56. P. 1022.
267. Boswell R.W., Morey I.J., J. Appl. Phys. Lett. 1988. V. 52. P. 21.
268. Godyak V.A., Piejak R. B., Phys. Rev. Lett. 1990. V. 65. P. 996.
269. Kazantsev S.A., Subbotenko A. V., J. Phys. D: Appl. Phys. 1987. V. 20. P. 741.
270. Volkova LM., Devyatov AM., Kral'kina E.A., Shushurin S.F., Vestn. Mosk. un-ta. Ser. Fizika, astronomiya. 1975. No. 6. P. 735.
271. Lowler J. E., Den Hartog, E., Hitchon, W., Phys. Rev. A. 1991. V. 43. P. 4427.
272. Wagner S.D., Ignatyev B.K., Tsendin L.D., Zh. Teor. Fiz. 1978. V. 48. P. 1191.
273. Medicus G.J., Appl. Phys. 1956. V. 27. P. 352.
274. Ionov N.I., Tondegode V.I., Zh. Teor. Fiz. 1964. V. 34. P. 354.
275. Savinov V.P., Singaevsky I.F., Proc. II Intern. scientific and technical. Conf. Problems and applied problems of physics. Saransk, 1999. P. 10.
276. Karkhov A.N., Zh. Eksp. Teor. Fiz. 1970. V. 59. P. 356.
277. Ginzburg V.L. The propagation of electromagnetic waves in a plasma. - Moscow: Fizmatgiz, 1960.
278. Kaptsov N.A., Electrical phenomena in gases and vacuum. - Moscow: GITTL, 1950.
279. Aleksandrov A.F., Kovalevsky V.L., Savinov V.P., Singayevskii, Proc. VIII Conf. on the physics of gas. discharge. Ryazan, 1996. Part 2. P. 48.
280. Aleksandrov A.F., Savinov V.P., Singaevsky I.F,. Proc. II Intern. scientific and technical. Conf. Problems and applied problems of physics. Saransk, 1997. P. 36.
281. Alexandrov A.F., Savinov V.P., Singaevsky I. F., Contr. Papers of ICPIGXXIII. Toulouse, 1997. V. 1. P. 108.
282. Aleksandrov A.F., Savinov V.P., Singaevsky I.F., Proc. IX Conf. on the physics of gas. discharge. Ryazan, 1998. Part 1. P. 100.
283. Savinov V.P., Singaevsky I.F., Proc. IV All-Russia. Scientific Conf. Educational experiment in higher education. Penza, 1998. P. 46.
284. Aleksandrov A.F., Savinov V.P., Singaevsky I.F., Vestn. Mosk. un-ta. Ser. Fizika, astronomiya. 1998. № 6. P. 52.
285. Kovalevsky V.L., Savinov V.P., Singaevsky I.F., Proc. Conf. Problems and applied problems of physics. Saransk, 1990. P. 3.
286. Aleksandrov A.F., Savinov V.P., Singaevsky I.F., Proc. Conf. Problems and applied problems of physics. Saransk, 1999. P. 9.
287. Aleksandrov A.F., Savinov V.P., Singaevsky I.F., Proc. All-Russian. Conf. Physical

electronics. Makhachkala, 1999. P. 50.

288. Aleksandrov A.F., Savinov V.P., Singaevsky I.F., Izv. RAS. Ser. fiz. 2000. v~. 64. P. 1387.

289. Aleksandrov A.F., Kovalevsky V.L., Savinov V.P., Singayevsky I.F., Proc. X Conf. on the Physics of gas. discharge. Ryazan, 2000. Part 2. P. 158.

290. Kovalevsky V.L., Savinov V.P., Singaevsky IF., Yakunin VG., Proc. IX Conf. on the Physics of gas. discharge. Ryazan, 1998. Part 1. P. 102.

291. Kovalevsky V.L., Savinov V.P., Singaevsky I.F., Izv. RAS. Ser. Fiz. 2000. V. 64. P. 1363.

292. O'Connell D., Turner M.M., Ellingboe A.R., Bull. Amer. Phys. Soc. 2004. V. 49. P. 27.

293. Paterson A., Panagopoulos T., Todorow V., et al., Bull. Amer. Phys. Soc. 2004. V. 49. P. 50.

294. Lee S.H., Tiwari P. K., Lee J.K., Plasma Sources Sci. Technol. 2009. V. 18. P. 025024.

295. Savinov V.P., Kruglov M.S., Yakunin V.G., Uchebnyi eksperiment v obrazovanii. 2013. No.3. P. 73.

296. Klykov I.L., Tarakanov V.P., Shustin E.G., Fizika plazmy. 2012. V. 38, No. 3. P. 290.

297. Schveigert I.U., Piters F.M., Pis'ma Zh. Teor. Fiz. 2007. V. 86 No. 9. P. 662

Index

R

radius
  Debye radius 86, 117, 176, 184
  Larmor radius 200, 205, 206, 214
Ramsauer gas 259, 270
RF detection 73, 83, 89, 90, 93, 95, 96, 97, 99, 112, 164, 183, 199, 250
Rogowski belt 24, 25

S

self-bias 90
space
  Aston dark space 122
  Crookes dark space 122, 190
  Faraday dark space 34, 41, 122
structure
  spatial structure 34, 37, 39, 42, 48, 61, 72, 73, 86, 121, 138, 229, 309

T

transition
  α → γ-transition 39, 40, 60

V

velocity
  drift velocity of the electron 6
  drift velocity of the electrons 17, 18

W

wall electron beams 85, 183, 259

9 780367 571627